Human Genome Methods

Human Genome Methods

Edited by

Kenneth W. Adolph, Ph.D.

Department of Biochemistry
University of Minnesota Medical School
Minneapolis, Minnesota

CRC Press
Boca Raton   New York

Acquiring Editor: Paul Petralia
Project Editor: Sarah Fortener
Marketing Manager: Becky McEldowney
Cover Designer: Dawn Boyd
Manufacturing: Carol Royal

Library of Congress Cataloging-in-Publication Data

Human genome methods / edited by Kenneth W. Adolph.
     p.   cm.
     Includes bibliographical references and index.
     ISBN 0-8493-4411-5 (alk. paper)
     1. Human genetics--Laboratory manuals.   2. Molecular biology--Laboratory
manuals.   3. Genetic engineering--Laboratory manuals.   4. Human gene mapping--Laboratory
manuals.   I. Adolph, Kenneth W., 1944–
     QH431.H8368 1997
     599.93′5—dc21   97-36261
                                                                                           CIP

Preface

Human Genome Methods is a practical guide to the application of important molecular biology and genetics techniques to research on human cells. The chapters are written by experts who are recognized authorities in their research areas and often originated the new techniques that are described. The central parts of the chapters are the experimental protocols which are presented so as to be readily used at the laboratory bench. These step-by-step protocols are intended to be concise and easy to follow. The procedures should be reproducible in the hands of researchers with varying levels of expertise. Suggestions to apply the procedures successfully are included, along with recommended materials and suppliers. A special feature of the chapters is that, in addition to the protocols, important background information and significant results of applying the methods are given. The methods are described in relation to specific research problems, but it should be emphasized that the experimental procedures should be easy to adapt to different research questions.

Molecular biology and genetics approaches now pervade research on human cells. A working knowledge of these gene-centered approaches is therefore essential. DNA sequencing, cloning, and related techniques are basic to investigations of the structure and function of human proteins, since these investigations are founded upon, or are incomplete without, knowledge of gene structure and regulation. Similarly, cell and developmental biology studies involving human cells are focused on molecular biology and genetics approaches. *Human Genome Methods,* therefore, aims to be a useful and convenient source of some of the newest molecular techniques of research on human cells.

The first section of the book, DNA Analysis, covers topics including microsatellite DNA, dynamic mutations, gene targeting using the DNA triple helix, and protease footprinting of DNA-protein interactions. This is followed by the Gene Expression section, which includes discussion of *in situ* hybridization and also cell synchronization and cell cycle-specific gene expression. The Programmed Cell Death; Cell

Culture section covers the emerging research area of programmed cell death (apoptosis) and the culture and analysis of cancer cells. Next to receive attention in the Transgenes, Knockout Mice, and Mouse Models section are methods related to transgene analysis of mouse embryonic stem cells, generation and knockout studies with null mutant mice, and mouse models for human disease. The final section, Genome Mapping, emphasizes the construction of linkage maps and somatic cell hybrids for mapping disease genes.

The volume is intended for researchers using molecular biology and genetics techniques to investigate human cells. The laboratory guide will therefore be valuable for biomedical researchers in fields such as human genetics, biology, biochemistry, cell biology, pathology, and cancer research. Established investigators, postdoctoral researchers, graduate students, and technicians will all find the work to be useful.

Kenneth W. Adolph

The Editor

Kenneth W. Adolph, Ph.D., is currently Associate Professor in the Department of Biochemistry of the University of Minnesota Medical School in Minneapolis. He received his Ph.D. from the Department of Biophysics of the University of Chicago. His B.S. and M.S. degrees were earned from the University of Wisconsin-Milwaukee. After receiving the Ph.D., he was a postdoctoral fellow at the Medical Research Council Laboratory of Molecular Biology in Cambridge, England, and at the Rosenstiel Basic Medical Sciences Research Center, Brandeis University. He was then a postdoctoral fellow in the Department of Biochemical Sciences, Princeton University. More recently, Dr. Adolph was a visiting scientist in the Department of Biochemistry of the University of Washington in Seattle. He is a member of the American Society for Biochemistry and Molecular Biology.

Dr. Adolph has research interests in gene structure and regulation, chromosome organization, nonhistone chromosomal proteins, and virus assembly. Research projects have been concerned with the structure and regulation of the human thrombospondin genes, electron microscopy studies of chromosome organization through the cell division cycle, the synthesis and post-translational modifications of nonhistone chromosomal proteins, and the assembly and structure of plant viruses.

Contributors

Urs Albrecht, Ph.D.
Research Associate
Department of Biochemistry
Baylor College of Medicine
Houston, Texas

Constanze Bonifer, Ph.D.
Assistant Professor
Institute for Biology III
Albert-Ludwigs Universität Freiburg
Freiburg, Germany

Mark Bouzyk, B.Sc.
Department of Genetic Technologies
Biopharmaceutical Research
 and Development
SmithKline Beecham
Essex, England

Ulrica Brunsberg, Ph.D.
Section for Medical Inflammation
 Research
Lund University
Lund, Sweden

Michael Carey, Ph.D.
Associate Professor
Department of Biological Chemistry
Center for Health Sciences
University of California-Los Angeles
School of Medicine
Los Angeles, California

Stefan Carlsén, Ph.D.
Section for Medical Inflammation
 Research
Lund University
Lund, Sweden

A. Collins, Ph.D.
Human Genetics
University of Southampton
Princess Anne Hospital
Southampton, England

Steven F. Dowdy, Ph.D.
Assistant Investigator
Howard Hughes Medical Institute
Division of Molecular Oncology
Washington University School
 of Medicine
St. Louis, Missouri

Gregor Eichele, Ph.D.
Professor
Department of Biochemistry
Baylor College of Medicine
Houston, Texas

Nicole Faust, Ph.D.
Research Assistant
Institute for Biology III
Albert-Ludwigs Universität Freiburg
Freiburg, Germany

Peter M. Glazer, M.D., Ph.D.
Associate Professor
Departments of Therapeutic
   Radiology and Genetics
Yale University School of Medicine
New Haven, Connecticut

John Groffen, Ph.D.
Professor of Research Pathology
   and Microbiology
Department of Pathology
Children's Hospital of Los Angeles
   Research Institute
Los Angeles, California

Rajagopal Gururajan, Ph.D.
Research Associate
Department of Tumor Cell Biology
St. Jude Children's Research Hospital
Memphis, Tennessee

Ann-Sofie Hansson, M.D.
Section for Medical Inflammation
   Research
Lund University
Lund, Sweden

Leena Haataja, Ph.D.
Department of Pathology
Children's Hospital of Los Angeles
   Research Institute
Los Angeles, California

Nora Heisterkamp, Ph.D.
Professor of Research Pathology
   and Microbiology
Department of Pathology
Children's Hospital of Los Angeles
   Research Institute
Los Angeles, California

Elizabeth Petri Henske, M.D.
Associate Member
Fox Chase Cancer Center
Philadelphia, Pennsylvania

Rikard Holmdahl, M.D., Ph.D.
Professor
Section for Medical Inflammation
   Research
Lund University
Lund, Sweden

Roderick Hori, Ph.D.
Postdoctoral Fellow
Department of Biological Chemistry
Center for Health Sciences
University of California-Los Angeles
School of Medicine
Los Angeles, California

Liselotte Jansson, Ph.D.
Assistant Professor
Section for Medical Inflammation
   Research
Lund University
Lund, Sweden

David F. Jarrard, M.D.
Assistant Professor of Surgery
Department of Surgery
University of Wisconsin Medical
   School
Madison, Wisconsin

Vesa Kaartinen, Ph.D.
Assistant Professor of
  Research Pathology
Department of Pathology
Children's Hospital of Los Angeles
  Research Institute
Los Angeles, California

David P. Kelsell, Ph.D.
Department of Genetic Technologies
Biopharmaceutical Research
  and Development
SmithKline Beecham
Essex, England

Vincent J. Kidd, Ph.D.
Associate Member
Department of Tumor Cell Biology
St. Jude Children's Research Hospital
Memphis, Tennessee

Susanne König, Dipl. Biol.
Institute for Biology III
Albert-Ludwigs Universität Freiburg
Freiburg, Germany

R. Frank Kooy, Ph.D.
Department of Medical Genetics
University of Antwerp
Antwerp, Belgium

David J. Kwiatkowski, M.D., Ph.D.
Associate Professor of Medicine
Experimental Medicine Division
Brigham and Women's Hospital
Boston, Massachusetts

Jill M. Lahti, Ph.D.
Assistant Member
Department of Tumor Cell Biology
St. Jude Children's Research Hospital
Memphis, Tennessee

Reuben Lotan, Ph.D.
Professor
Department of Tumor Biology
The University of Texas
  M.D. Anderson Cancer Center
Houston, Texas

Hui-Chen Lu, B.S.
Research Assistant
Department of Biochemistry
Baylor College of Medicine
Houston, Texas

Anna Mikulowska, D. Med. Sci.
Section for Medical Inflammation
  Research
Lund University
Lund, Sweden

N. E. Morton, Ph.D.
Professor
Human Genetics
University of Southampton
Princess Anne Hospital
Southampton, England

Ben A. Oostra, Ph.D.
Professor of Human Genetics
Department of Clinical Genetics
Erasmus University
Rotterdam, The Netherlands

Joan Overhauser, Ph.D.
Associate Professor
Department of Biochemistry
  and Molecular Pharmacology
Jefferson Medical College
Philadelphia, Pennsylvania

Ulf Pettersson, M.D., Ph.D.
Professor
Department of Medical Genetics
Uppsala University
Uppsala, Sweden

Jean-Pierre Revelli, Ph.D.
Research Associate
Department of Biochemistry
Baylor College of Medicine
Houston, Texas

Catherine A. Reznikoff, Ph.D.
Professor
Department of Human Oncology
Program in Cellular and
 Molecular Biology
Comprehensive Cancer Center
University of Wisconsin
 Medical School
Madison, Wisconsin

Greg H. Schreiber, B.S.
Howard Hughes Medical Institute
Division of Molecular Oncology
Washington University School
 of Medicine
St. Louis, Missouri

Albrecht E. Sippel, Ph.D.
Professor for Genetics
Institute for Biology III
Albert-Ludwigs Universität Freiburg
Freiburg, Germany

Nigel K. Spurr, Ph.D.
Department of Genetic Technologies
Biopharmaceutical Research
 and Development
SmithKline Beecham
Essex, England

Mats Sundvall, D. Med. Sc.
Department of Medical Genetics
Uppsala University
Uppsala, Sweden

Linda F. VanDyk, Ph.D.
Howard Hughes Medical Institute
Division of Molecular Oncology
Washington University School
 of Medicine
St. Louis, Missouri

Mikael Vestberg, M.Sc.
Section for Medical Inflammation
 Research
Lund University
Lund, Sweden

Gan Wang, Ph.D.
Postdoctoral Associate
Departments of Therapeutic
 Radiology and Genetics
Yale University School of Medicine
New Haven, Connecticut

Patrick J. Willems, Ph.D., M.D.
Professor of Medical Genetics
Department of Medical Genetics
University of Antwerp
Antwerp, Belgium

Xiao-Chun Xu, M.D., Ph.D.
Assistant Professor
Department of Clinical
 Cancer Prevention
The University of Texas
 M.D. Anderson Cancer Center
Houston, Texas

Thomas R. Yeager, Ph.D.
Postdoctoral Fellow
Department of Human Oncology
Program in Cellular and
 Molecular Biology
Comprehensive Cancer Center
University of Wisconsin
 Medical School
Madison, Wisconsin

Contents

Section III. Programmed Cell Death; Cell Culture

Section IV. Transgenes, Knockout Mice, and Mouse Models

Section V. Genome Mapping

Section I

DNA Analysis

Chapter

Human Microsatellite Repeat Markers and Their Application to Analysis of Clonality and Allelic Loss in Tumors

Elizabeth Petri Henske and David J. Kwiatkowski

Contents

0-8493-4411-5/98/$0.00+$.50
© 1998 by CRC Press LLC

I.   Human Microsatellites:
Frequency and Polymorphism

Tandemly repeated sequences for which the core sequence has a length of one to six nucleotides are common in eukaryotic genomes. These repeats have been called simple tandem repeats (STR), simple sequence repeats (SSR), and microsatellites; here, we use the term microsatellites. Because their length is frequently polymorphic, microsatellites are useful for genetic mapping, assessment of clonality, and analysis of tumors and pre-neoplastic lesions for allelic loss.

A.   Dinucleotide Repeats

All eukaryotic species have repeats of the simple sequence (dC:dA)n or (dG:dT)n, which will be designated (GT)n in this chapter. These repeats do not occur in bacteria.[1] The number of (GT)n repeats differs among eukaryotic species, ranging from 100 copies in yeast to 200,000 copies in salmon.[2]

The human genome is estimated to contain 50,000 (GT)n, where n ≥ 9. This would put a (GT)n repeat every 60 kb in the genome, if uniformly spaced. (GT)n repeats have been shown to occur every 30 kb in the euchromatic regions of human chromosome 16, with under-representation in heterochromatin.[3] Direct sequence analysis of 685 kb of the human T-cell receptor beta-chain gene complex has shown that dinucleotide repeats of length n ≥ 10 occur once every 38 kb, and (GT)n repeats of length n ≥ 10 occur every 76 kb.[4] (GT)n repeats are significantly more common in the rat and mouse genomes, occurring approximately every 20 kb.[3]

The functional significance of (GT)n repeats, which are found primarily within introns or between genes,[5] is not known. Roles in gene regulation[6] and recombination[7,8] have been proposed. Although the position of the repeats tends to be conserved between closely related species such as man and monkey or mouse and rat, they are usually not conserved between distantly related species such as rodent and human.[3]

(GT)n repeats frequently exhibit length polymorphism.[5] Most often, alleles of different sizes vary in length by multiples of two bases, due to the presence of variable numbers of the (GT)n repeat. The frequency of this phenomenon has been shown to depend upon both the length of the repeat and whether there are internal variations in the repeat sequence.[9,10] A convenient and simple measure of the polymorphism of a microsatellite marker is the frequency of heterozygosity (two different alleles) seen in a reference population.

As the length of a (GT)n repeat increases, both the frequency of heterozygosity and the level of heterozygosity increase. The heterozygosity of imperfect repeats depends most strongly on the length of the longest uninterrupted repeat. Repeats of length 10 or less are rarely polymorphic to any useful degree; conversely over 80% of repeats of length 16 or greater have a heterozygosity rate of ≥ 50%.[9,10] Of repeats of length ≥ 20, 50 to 75% have heterozygosity ≥ 70%. Analysis of the 21 dinucleotide repeats described above in the T-cell receptor beta locus yielded 14 polymorphic markers, or one on average every 49 kb, with heterozygosities between 0.35 and 0.82.[4]

B. Mononucleotide Repeats

The frequency and heterozygosity of mononucleotide repeat sequences in the genome have not yet been investigated in detail. There are several cases described, however, in which the mononucleotide repeat (A)n is polymorphic.[11] The problem with using these markers is that (A)n repeats are found almost exclusively as the poly A tail of Alu repeats. This creates technical problems with primer design and analysis (see further discussion below), so that, although potentially useful, such repeats are best avoided. In addition, since mononucleotide repeats vary by single bases, interpretation of adjacent alleles is very difficult, and the stutter pattern is extensive.

C.   Trinucleotide Repeats

Trinucleotide repeats occur less frequently than dinucleotide repeats. There are 10 possible combinations of three nucleotides. Of these 10 classes, repeats of the sequence AAT are most frequent and useful for development of polymorphisms, with an estimated frequency of 1 per 510 kb of genomic DNA, or an estimated 5900 copies in the human genome.[12] Other trinucleotide repeats of potential usefulness as markers, in declining order of utility, are ACT, AAG, ATC, and AAC.

As with (GT)n repeats, the frequency of polymorphism of a trinucleotide repeat depends upon the length of the repeat. Of trinucleotide repeats with length $\geq$ 9 repeats, 90% have heterozygosity greater than 50%, and about 70% of repeats with length $\geq$ 13 have heterozygosity greater than 70%.[12]

Trinucleotide sequences that are A/T rich (AAC, AAG, AAT) are often found adjacent to Alu repeats, limiting their utility. Information on ACG and CCG repeat sequences is not available because these repetitive sequences were found to be difficult to passage in bacteria.

Analysis of the 685-kb T-cell receptor beta locus yielded two trinucleotide polymorphic markers with heterozygosities of 0.57 and 0.67, and two shorter trinucleotide repeats (length of 8 repeats) which were not evaluated for heterozygosity.[4]

D.   Tetranucleotide Repeats

There are 30 possible tetranucleotide repeat sequences. This large panel of potential polymorphisms has been investigated by two research groups to determine their potential utility as markers, resulting in the identification of many hundreds of markers of this type at this time.[13,14] Core sequences of the forms GATA, GAAA, GGAA, GGAT, GAAT, AAAT, GTAT (in declining order of utility) have been found to be most useful for markers. About two thirds of the currently identified markers are GATA repeats.

The genome frequency of tetranucleotide markers is estimated to be one marker per 200 to 1000 kb of genomic DNA. Because tetranucleotide repeats are more frequently associated with Alu repetitive elements than GT repeats, only a fraction of them with length greater than 7 is suitable for the development of a robust polymorphic marker.

Analysis of the 685-kb T-cell receptor beta locus yielded three tetranucleotide polymorphic markers with heterozygosities of 0.23 to 0.67.[4] Seventeen other tetranucleotide sequences were of short length (4 to 8 repeats) and therefore were not evaluated for heterozygosity.

II.   Methods for Identification and Development of Microsatellites from a Genomic Clone

A.   General Considerations and Strategy

When the entire sequence of the human genome is known, it will be an easy matter to identify potentially polymorphic markers by examination of the primary sequence and consideration of the information presented above. In addition, by then it is likely that the majority of all polymorphic microsatellites already will have been discovered by the large-scale efforts which have been so productive thus far. Therefore, a first step in identification of a polymorphism from a given genomic clone is to examine the appropriate databases (GDB, Genethon, and CHLC web sites) to identify known microsatellites which are from the chromosomal location of the clone. If this method fails to be productive, then the following approach to identification of a polymorphic microsatellite may be used.

B.   Determination that a Genomic Clone Contains a Likely Microsatellite for Primer Preparation

For the identification of GT repeats in a genomic clone, we have found the following method to be effective because polymorphic GT repeats are reliably identified, and shorter repeats (which are unlikely to be polymorphic) are avoided. Hybridization to colony lift filters or Southern blot analysis of the genomic clone DNA is performed with a GT oligonucleotide probe of length 33 labeled with ^{32}PγATP by polynucleotide kinase. Hybridization is carried out at 65°C with washing at the same temperature in 1% SDS, 40 mM NaPO$_4$ (pH 7.2), and 1 mM EDTA.

For identification of tri- and tetranucleotide microsatellites, we have used kinase-labeled oligonucleotides for the following sequences: AAT, GATA, GAAA, GATT, GAAT, GGAT, and GGAA (each of length 40 bases). These microsatellite sequences were selected out of the larger pool of all tri- and tetra-nucleotide sequences based upon their frequency in the genome and the frequency with which adjacent Alu repeats interfere with primer development.

These seven oligonucleotides require individualized hybridization and wash conditions to ensure identification of polymorphic microsatellites and avoidance of shorter repeats (Table 1). Using these hybridization conditions for (GT)n, GATA, and ATA repeats, we have found that approximately 80% of identified sequences are polymorphic. The remaining 20% that are not polymorphic generally have been imperfect (interrupted) GT repeats or shorter repeats of borderline length but have also included some long microsatellites — (GT)n repeats up to 25 in length — which are not polymorphic.

From a comprehensive analysis of 1.5 Mb of human 9q34 using these methods to identify potential microsatellites, we found 15 (GT)n, 2 AAT, and no tetranucleotide

TABLE 1.1
Microsatellite Hybridization
Temperatures

Sequence	Hybridization and Washing Temperatures (°C)[a]
AAT	37
GATA	45
GAAT	58
GAAA	58
GATT	58
GGAA	65
GGAT	65

[a] Hybridizations are carried out in 0.5 M NaPO$_4$, pH 7.2; 7%
SDS; 1% BSA. Washes are performed in 40 mM NaPO$_4$,
pH 7.2; 1% SDS; 1 mM EDTA.

microsatellites. This frequency is in accord with the genome wide-frequency calcu-
lations presented above.

C.   Determination of Flanking Sequence Surrounding a Microsatellite for Primer Preparation

Several methods have been described for efficient isolation of microsatellite elements from genomic clones of cosmid and larger size.[10,15,16] The most reliable method
in our experience is preparation of a plasmid library from the genomic clone,
followed by screening with the appropriate microsatellite probe by the filter-based
colony lift technique. For this purpose, we have typically used the restriction enzyme
*Pst*I, since it provides the best size range for subfragments, with most fragments
being between 0.5 and 2 kb in size. These size fragments provide a balance between
obtaining an adequate flanking sequence for primer selection and minimal sequencing analysis.

It is also possible to digest the genomic clone with several restriction enzymes
or combinations, followed by Southern blot analysis with the microsatellite probe to
identify fragments containing the microsatellites that are of a size appropriate for
sequence analysis, followed by specific subcloning.

When a plasmid subclone is available, sequence information may be obtained
from both ends of the insert by standard methods (Sequenase, or Taq polymerase

cycle sequencing). If the microsatellite sequence is neither encountered nor visualized on the autorad, then the size of the insert should be reviewed to ensure that the clone indeed contains the microsatellite sequence.

If the microsatellite sequence is completely readable, then flanking primers may be chosen. If the sequence leading into the microsatellite is readable but the other side is not readable, then a single flanking primer may be made and used in a subsequent sequencing reaction to determine the sequence on the opposite side.

If the clone insert is of a size greater than 1.5 kb, and the microsatellite is not seen, there are two possible strategies. The first is to digest the clone insert with several restriction enzymes and identify a smaller fragment that contains the microsatellite. This is most easily done by generating a deletion construct, using an enzyme that digests within both the polylinker region of the vector and the genomic insert so that digestion, self-ligation, and introduction into competent cells may be performed efficiently.

A second method, for determination of (GT)n flanking sequences, is to use the oligonucleotides (GT)9X, where X is A, C, or T, and (TG)9Y, where Y is A, G, or C, as sequencing primers. These oligonucleotides will anneal specifically at the end of a (GT)n microsatellite, giving interpretable sequence data most of the time for plasmid clones with inserts < 5 kb. They will also give a flanking sequence from a cosmid clone about 25% of the time and thus could be attempted earlier in this process.

D.   Primer Selection

Primer pairs flanking the microsatellite repeat can be designed using the PRIMER program. Although values from 50 to 75°C can be used, we prefer to design primers with a melting temperature (T_m) of 60°C. It is essential that the primers have nearly identical melting temperatures and that they have a minimum length of 18 bases.

We prefer primers close to (but not within) the repeat. There is no advantage to larger amplification products. Shorter products mean shorter gel run times for analysis and better efficiency in amplification of DNA extracted from paraffin-embedded tissue (see Part V of this chapter).

Often one or another flanking sequence for a microsatellite consists of an Alu repetitive element. As discussed above, this is a much greater problem with tetranucleotide repeats than with GT dinucleotide repeats. When an Alu is encountered, the only choice is to design a primer which is maximally divergent from the Alu consensus sequence. This is usually possible. The microsatellite flanking sequence is compared with the Alu consensus to identify regions of maximal divergence. A primer is chosen which has approximate T_m and maximal divergence, optimally three or more bases from the consensus. A Blast comparison is then performed with the proposed oligonucleotide to determine the number of sequences in the human genome that match with the oligo and the degree of the match. If fewer than 10 sequences in the genome (from the current Genbank collection) match at 19 or 20 positions within the oligonucleotide, then the primer is synthesized and tested. In our experience, most of the time this method effectively identifies useful primers that work in PCR amplification to give an interpretable signal, but often with some background amplification.

E.   Alternative Strategies

For large-scale preparation of microsatellites of a particular class, two methods of enrichment for such sequences have been described, both of which depend upon hybridization of genomic DNA to the target core sequence. These methods are robust and efficient for large-scale preparations but are unsuitable for routine use in microsatellite development from a single genomic clone.[17,18]

III.   Methods for Analysis of Microsatellites

(GT)n or trinucleotide polymorphisms are analyzed by amplifying the segment of DNA that contains the repeat using the polymerase chain reaction (PCR). Several methods may be used for this procedure. For small-scale situations (<100 PCR reactions), we use the following protocol.

1.   Prepare the following reaction mix, scaled up for the number of reactions to be performed:

> 2 μl dNTP mix (1 mM dATP, dCTP, dTTP; 50 μM dGTP)
> 0.4 μl oligonucleotide primer pair mix (100 ng/μl, ea)
> 0.027 μl α^{32}P-dGTP (NEN NEG-014H, 3000 Ci/mM)
> 1 μl 10× reaction buffer (0.5 M KCL; 0.1% gelatin; 0.2 M tris-HCl, pH 8.5;
>   15 mM MgCl$_2$)
> 0.027 μl Taq polymerase (Perkin-Elmer Cetus)
> 1.546 μl H$_2$O

> 5 μl total volume

2.   Aliquot 5 μl of genomic DNA (1 to 10 ng/μl, usually use 1:100 dilution of standard WBC DNA prep) into 0.5 ml Eppendorf tubes or into a 96-well tray. Add 5 μl of the reaction mix prepared in step 1.

3.   We prefer to use a heated lid machine, with firmly capped tubes and no oil. If using oil, cover the mix in each tube with about 20 μl of light mineral oil (1 drop).

4.   Place samples in thermal controller with the following settings:
       a. 94°C for 1.5 min
       b. 94°C for 30 sec
       c. 55°C for 30 sec (adjust for Tmax of oligos)
       d. 72°C for 40 sec (increase time if product length > 300 bp)

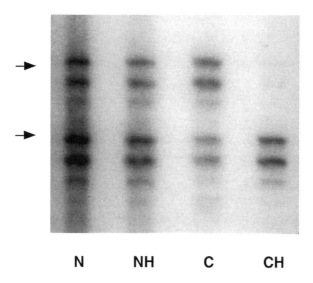

N   NH   C   CH

FIGURE 1.1

Example of clonality detected using the HUMARA assay in epithelial cells from a renal cyst. Two alleles, each represented by two bands (the uppermost of which is indicated with an arrow) are present in normal undigested DNA (N), normal DNA digested with *Hpa*II (NH), and undigested cyst epithelial cell DNA (C). The upper allele is lost in *Hpa*II-digested cyst epithelial cell DNA (CH), indicating clonality.

    e. Repeat steps b, c, and d 29 to 34 times

    f. 72°C for 4 min, 20 sec

5.   Add 10 µl of loading buffer (98% formamide/bromphenol blue-xylene cyanol/10 m*M* EDTA); 5 µl only for the oil-less method.

6.   Heat samples at 80 to 100°C for 5 min and load on a 6% sequencing gel. If using the oil-less PCR method, a longer heating time can be used to concentrate samples up to fourfold in order to obtain a more intense signal.

      This protocol, which we have found to be robust and simple, uses so-called internal labeling in which the label is incorporated by the Taq polymerase into all amplified products. Several alternatives are possible. Many investigators feel that using a radiolabeled oligonucleotide to label the reaction products provides a distinct advantage in interpretation of the autoradiographic signal. In our experience, this advantage is minimal. A few investigators prefer to analyze the products of the PCR reaction on nondenaturing polyacrylamide gels and feel that the autoradiographic signal is again easier to interpret than on ordinary denaturing gels.

      PCR products analyzed by denaturing gel eletrophoresis always show some degree of "stutter" bands. That is, a single band is not seen per individual allele (Figures 1.1 to 1.3). This phenomenon has at least three causes: two products are always derived from each allele (both strands); the Tdt activity of Taq polymerase adds an additional base variable to each strand; and, most importantly, the apparent ability of the extending strand to "stutter" or move forward or backward one repeat unit in its apposition to the opposing strand. Thus, interpretation of alleles often

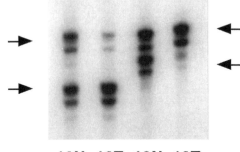

18N 18T 12N 12T

FIGURE 1.2
PCR products using the chromosome 16p13 dinucleotide repeat marker Kg8 from two sporadic angiomyolipomas (18T and 12T) and corresponding normal tissue (18N and 12N). Each allele is represented by a primary upper band (arrow) and a less prominent lower band. Loss of the upper allele in 18T and loss of the lower allele in 12T are seen. (From Henske, E. et al., *Genes Chrom. Cancer*, 13, 295, 1995. With permission.)

requires experience. For the first-time user, it is very helpful to have a simple, two-generation pedigree to analyze to confirm interpretation. Stutter bands are much less prominent for tri-, tetra-, and higher order repeat units and are much worse for mononucleotide repeats than for dinucleotide repeats. This observation has spurred significant efforts in the identification of tri- and tetranucleotide repeats.

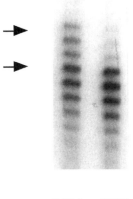

33N   33T

FIGURE 1.3
PCR products using the chromosomes 16p13 dinucleotide repeat marker D165525 from a tuberous sclerosis-associated angiomyolipoma (33T) and corresponding normal tissue (33N). Each allele is represented by three bands of approximately equal intensity, the uppermost of which is indicated with an arrow. Loss of the upper allele is seen in 33T. (From Henske, E. et al., *Genes Chrom. Cancer,* 13, 295, 1995. With permission.)

When large numbers of PCR reactions must be analyzed on a set of genomic DNA samples, there are several procedures to improve the efficiency of the analysis. One option is multiplexing, or the incorporation of several pairs of PCR primers in a single reaction tube. When performed with internal labeling, however, this becomes difficult with even two or three primer pairs due to overlapping autoradiographic signals and background.

An alternative method used widely at Genethon and elsewhere is to perform the PCR reactions without label. The PCR products are then transferred to a nylon filter and probed with an oligonucleotide that is specific to a single amplified product. By serial probing with distinct oligos of the same nylon filter, clear patterns for each individual PCR reaction can be determined. Since the gel run for separation is the rate-limiting step in this procedure, this multiplexing procedure is valuable.

Finally, when a fluorescent automated sequencer system is available, the products may be labeled by attaching fluorophores to one of the oligos used in the reaction. By this means, multiplexing is accomplished for both the PCR reaction and the gel run; however, interpretation of the output signal is not a routine task for any of the automated sequencer systems. Fortunately, improvements in analytic software are anticipated. This problem is reduced for tetranucleotide and, to a lesser extent, for trinucleotide repeats, as the pattern of redundant bands is simpler.

IV. Clonality Analysis Using the Androgen Receptor Short Tandem Repeat

A. The HUMARA Clonality Assay

The HUMARA PCR-based clonality assay takes advantage of a highly polymorphic trinucleotide repeat in the first exon of the human androgen receptor (HUMARA) gene.[19] The polymorphic repeat in the HUMARA gene allows the maternally and paternally derived copies of X to be distinguished in heterozygous females.

In females, one X chromosome or the other is inactivated in all cells early in embryonic development (Lyonization), through a process in which most genes on the inactive X chromosome are methylated at multiple CG sites. Such methylation of the inactive X occurs within the HUMARA locus at multiple sites, including two *Hpa*II restriction enzyme sites (CCGG) located 100 bp 5' to the repeat (Figure 1.4). Since only the inactive copy of X is methylated and *Hpa*II does not cut methylated DNA, the *Hpa*II site allows the active and inactive copies of X to be distinguished.

In the HUMARA clonality assay, normal DNA and DNA from a lesion are digested with *Hpa*II. When PCR is performed using primers that flank the *Hpa*II sites and the CAG repeat, as indicated in Figure 1.4, two alleles will be seen in undigested normal and tumor DNA from heterozygous females. In digested normal DNA, methylation of the X chromosome will have occurred randomly on the maternally

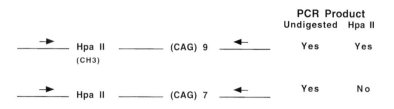

FIGURE 1.4

Relative positions of the *Hpa*II site, the CAG repeat, and oligonucleotide primers (indicated with arrows) used in the HUMARA clonality assay. In this example, the top copy of the X chromosome has an allele with nine CAG repeats and the lower copy of X has an allele with seven CAG repeats, allowing the two copies to be distinguished by size after PCR amplification of undigested DNA. The top copy is methylated and thereby protected from *Hpa*II digestion. In digested DNA, the top copy, therefore, will still amplify. The lower copy is unmethylated; it will be cut by *Hpa*II, and no PCR product will be produced after digestion.

and paternally derived copies of X, and both alleles of the CAG repeat will amplify. In a clonal tumor, however, all cells will have either the maternally or paternally derived copy of X methylated (Figure 1.1). The unmethylated copy of X will be cut at the *Hpa*II site. The methylated copy will not be cut. Only the CAG allele on the methylated copy of X will amplify in a clonal tumor.

Although it is possible to analyze clonality by assessment of methylation in a Southern blot assay, the HUMARA assay has two important advantages. First, the (CAG)n repeat in the androgen receptor gene has a much higher heterozygosity (90%) than the conventional RFLP/Southern blot polymorphisms HPRT (30% heterozygosity) and PGK (34% heterozygosity).[20] Second, very small amounts of DNA can be used. This assay was used to demonstrate that Langerhans'-cell histiocytosis (histiocytosis X) is a clonal disorder, rather than a polyclonal reactive disease.[21] In this study, the HUMARA locus was informative in all ten patients, in contrast to conventional assays that were informative in only one patient.

B.   Reaction Conditions

To perform the assay, 5 µl (approximately 1 µg) of DNA solution is digested in a 20-µl reaction using 10 units of *Hpa*II. The sample is heated to 95°C for 10 min to inactivate the enzyme. One µl of the digested DNA is used in a 10-µl PCR reaction (under the conditions described in Section III of this chapter) using the primers described by Allen et al.,[19] available from Research Genetics. Undigested tumor and normal DNA should be included as controls. We amplify this locus for 30 cycles, each comprising 40 sec at 90°C, 30 sec at 55°C, and 40 sec at 72°C, following an initial denaturation at 95°C for 90 sec. Products, which are approximately 280 bp in size, are resolved by denaturing 8 *M* urea polyacrylamide gel electrophoresis followed by autoradiography.

If the amount of background is high, the annealing temperature should be raised in 2° increments until the signal is more specific. When first using this assay, it should be performed on DNA obtained from a family to ensure correct interpretation of the alleles.

An example of a clonal sample using this assay is shown in Figure 1.1. Two alleles are seen in normal DNA and undigested sample DNA, and one is seen in digested sample DNA.

C.   Experimental Considerations and Potential Problems

1.   Skewed X inactivation

Although generally equal numbers of cells in the adult female have experienced inactivation of each X chromosome, occasionally most cells have the same inactivated X, so-called extreme Lyonization. This situation confounds analysis of clonality, since all tissues in the body, clonal or not, will tend to show selective inactivation of a single X. Recent studies suggest that unequal Lyonization occurs in 23% of normal females.[22] Because of this, normal tissue (ideally from the same organ as the tumor since unequal Lyonization can be organ specific) should be analyzed from each patient and compared with the tumor.

In cases of extreme skewing in the normal tissue, it may not be possible to assess clonality in the tumor. In other cases, if the skewing in the normal tissue is in the opposite configuration from the skewing in the tumor, clonality may still be determined. Quantitation of the signals with a phosphorimager may be useful in identifying clonal samples in the case of skewed Lyonization.[21]

2.   Incomplete DNA digestion

If the *Hpa*II digestion is incomplete, the HUMARA assay may falsely indicate that a sample is not clonal, when in fact it is clonal. To determine the completeness of the digestion, digested DNA can be compared with undigested DNA on an agarose gel. Digestion of the DNA with *Rsa*1 prior to *Hpa*II digestion may be helpful in obtaining complete digestion with *Hpa*II.

3.   Source of DNA

The best source of tissue for clonality analysis is fresh or frozen tissue from which DNA can be extracted using conventional techniques. Clonality analysis using paraffin-embedded tissue has been described,[23] but it is potentially problematic if the fixation technique alters the methylation status of the DNA or interferes with restriction enzyme digestion. Primers that amplify a smaller product are required for paraffin-embedded tissue (it is usually difficult to amplify products greater than 200 bp from paraffin-embedded tissue; the conventional HUMARA primers amplify a 280-bp product).

4.   Contamination of tumor DNA
        with normal DNA

Since the HUMARA assay is PCR based, contamination of the tumor DNA sample with normal DNA could preclude the detection of clonality. It is important that the tumor DNA sample contain as pure a population of tumor cells as possible.

V.   Analysis of Allelic Loss and Microsatellite Instability in Tumors

Analysis of tumor DNA with polymorphic microsatellites can be used to detect lost chromosomal regions or to detect changes in microsatellite alleles (which may indicate repair enzyme defects in the tumor, as discussed below).

A.   Analysis of Tumors for Loss of Heterozygosity Using Microsatellites

Detection of lost or deleted regions in tumor DNA (allelic loss analysis or loss of heterozygosity [LOH] analysis) is based on the Knudson tumor-suppressor gene, "two-hit" model. In this model,[24] tumors develop when both copies of a tumor suppressor gene are inactivated. The "second hit" is frequently the loss of a relatively large chromosomal region containing the gene. This chromosomal loss can be detected by Southern blot analysis of restriction fragment length polymorphisms or PCR analysis using microsatellite markers (Figures 1.2 and 1.3). LOH has also been found in premalignant oral lesions, suggesting a possible value in cancer risk assessment.[26]

    Consistent LOH (50 to 100% of all tumors examined) in a given genomic region is strongly suggestive of the existence of a tumor suppressor gene in the region. Observation of LOH in a genomic region is typically followed by efforts to identify the minimal, "critical" region of LOH, and then a search for inactivating mutations in the remaining copy of candidate tumor suppressor genes from the region in the patient's tumor DNA samples. This approach contributed to the cloning of the neurofibromatosis 2 gene[25] but is often of limited utility in positional cloning efforts because the "lost" chromosomal regions in malignant tumors tend to be large (for example, entire chromosome arms).

B.   Microsatellite Instability

In human germline DNA, changes in microsatellite allele size occur at a frequency of about one per 1000 meioses. Microsatellite allele shifts are also seen at a roughly similar frequency when peripheral blood-derived DNA is compared with a lymphoblastoid cell line derived DNA from the same individual.[27,28] When changes do occur, they most commonly result in a gain of allele size of one repeat unit.[2,29]

Certain tumors have been discovered to have frequent microsatellite allele size alterations and have been described as having "microsatellite instability" or "mismatch repair deficiency". In these tumors, alleles are present in tumor DNA that are not present in normal DNA. Microsatellite instability is particularly common in tumors from patients with an inherited defect in one of four mismatch repair-related genes: MSH2, MLH1, PMS1, or PMS2 (reviewed in Reference 30). Microsatellite instability has also been seen in a wide variety of sporadic tumors (for example, see References 31 and 32). Microsatellite alterations have also been detected in the serum of patients with head and neck cancer and lung cancer, presumably from circulating tumor cells.[33,34]

C.   Experimental Conditions for Analysis of Microsatellites in Tumor DNA

DNA isolated from fresh or frozen tumor samples works consistently and efficiently in microsatellite analysis. A normal tissue sample is also required for preparation of normal DNA from each patient. DNA may also be prepared from paraffin-embedded tissue. Such material has limitations in microsatellite analysis but provides the ability for direct identification of tumor vs. normal cells, by comparison with H- and E-stained sections.

We have used DNA from paraffin-embedded tissue extensively in our analysis of lesions from tuberous sclerosis patients,[37,38] as shown in Figures 1.2 and 1.3, using the methods that follow.

1. Carefully examine an H- and E-stained section of the tumor to distinguish the tumor from surrounding normal tissue. If the boundaries of the tumor are not discrete, it is important to exclude all normal tissue (even if some tumor is included), since normal DNA could preclude the detection of LOH.

2. Using an unstained 5-μm section, scrape away any excess paraffin and any tissue that is not desired using a clean razor blade. After wiping the blade, scrape the remaining tissue into a 1.5-ml Eppendorf tube containing 50 mM KCl, 10 mM Tris (8.3), 1.5-mM MgCl$_2$, 100 μg/ml bovine serum albumin, 45% Tween 20, 0.45% NP40, and 100 μg/ml proteinase K. The volume of buffer depends on the amount of tissue. For an area approximately 1×1 cm, use 100 μl of buffer; for an area 1.5×1.5 cm, use 200 μl. After incubation at 50°C for 18 hr, heat the sample for 7 min at 95°C to activate the proteinase K.

3. The PCR amplification is performed using a 2-μl aliquot of DNA in a 20-μl reaction. The exact PCR conditions will depend on the primers used. A typical reaction condition is an initial denaturation at 95°C for 90 sec followed by 30 cycles of 94°C for 40 sec, 55°C for 40 sec, and 72°C for 40 sec. We do the PCR in the presence of α^{32}P-dGTP, as described earlier.

4. PCR products are analyzed by running on an 8-M urea sequencing gel followed by overnight autoradiography. Examples of LOH detected in paraffin-embedded tumor tissue are shown in Figures 1.2 and 1.3.

5. We use visual interpretation to determine LOH. Phosphorimaging may also be used. At least a 50% reduction in signal intensity in the tumor is required. We find it very useful to have equal loadings of the normal and tumor tissue before interpreting a sample as having LOH. This usually requires reloading the samples once or twice.

If alleles are closely spaced, it is difficult or impossible to detect loss of the lower allele, since stutter bands from the upper allele may overlap with the bands from the lower allele. It is critical to compare normal and tumor DNA. In normal DNA, the intensity of alleles can differ by as much as 50%, with the upper allele being less intense, simply because smaller PCR products amplify more efficiently. This effect is most marked when the alleles are widely spaced and could appear to indicate LOH in tumor DNA if not compared with the pattern in normal DNA.

D.   Experimental Considerations and Potential Problems

1.   Size of product

Smaller products amplify more readily from DNA prepared from paraffin-embedded tissue. Products less than 100 bp amplify best; consistent amplification can usually be obtained with products between 100 and 150 bp and sometimes with products between 150 and 180 bp. If a microsatellite product is too large for paraffin-embedded tissue, new primers usually can be designed to amplify a smaller product (see Section II.D of this chapter).

2.   Annealing temperature

It may be necessary to individualize reaction conditions for paraffin-embedded tissue. We find that annealing temperatures that work well on other types of DNA preparations (buccal swabs or DNA prepared from fresh or frozen tissue) frequently are not ideal for paraffin-embedded tissue. We adjust the annealing temperature higher (if excessive background bands are present) or lower (if the signal is faint) by 2 to 5°C.

3.   Alternate microsatellites

Some microsatellites that amplify well from other DNA sources will not amplify consistently from paraffin-embedded tissue despite alterations in the reaction conditions. It may be necessary to find other sources of DNA or to use another nearby microsatellite.

4.   Sample half-life

DNA from paraffin-embedded tissue amplifies most consistently in the first month after preparation. Experiments should be planned so that all of the reactions on a particular sample are performed in the weeks immediately after preparation. Some samples will continue to amplify well for more than a year, but many will have a gradual loss of activity. Several extra slides should be kept on hand so that fresh DNA can be prepared when needed. We store these at room temperature. While on the slide, the DNA has a much longer half-life (at least 2 years).

5.   Sample dilution

We have obtained the best results by preparing a concentrated DNA solution (a 1×1-cm area of tissue is extracted in 100 µl of extraction buffer) and using a 1-to-10 dilution of this DNA for PCR (i.e., 2 µl of this DNA in a 20-µl PCR). Some samples amplify better if diluted so that 1 µl or 0.5 µl is used in a 20-µl PCR.

6.   Artifactual bands

Artifactual bands that are more prominent in the normal tissue than in the tumor tissue are problematic, since they may falsely suggest that LOH is present. This can be particularly difficult when the normal and tumor tissue are from different sources. For example, when the normal DNA is prepared from paraffin-embedded tissue, the reaction tends to be more robust for the peripheral blood DNA, and extra bands that do not represent alleles may be present. In these cases, reaction conditions may have to be different for the normal and tumor DNA, with a higher annealing temperature or a more dilute sample for the normal DNA. Obtaining samples from first-degree family members can be extremely useful to confirm Mendelian inheritance of the alleles.

7.   Contamination of tumor DNA with normal DNA

Contamination of the tumor DNA sample with normal DNA could preclude the detection of LOH. It is important that the tumor DNA sample contain as pure a population of tumor cells as possible. Unstained sections should be meticulously compared with H- and E-stained sections in order to identify appropriate areas. If necessary, DNA can be extracted from H- and E-stained slides. In our experience, however, these samples do not amplify as reliably as unstained sections.

References

1.   Gross, D. and Garrard, W., *Mol. Cell.Biol.,* 6, 3010, 1986.

2.   Hamada, H. and Kakunaga, T., *Nature,* 298, 396, 1982.

3.   Stallings, R., Ford, A., Nelson, D., Torney, D., Hildebrand, C., and Moyzis, R., *Genomics,* 10, 807, 1991.

4.   Charmley, P., Concannon, P., Hood, L., and Rowen, L., *Genomics,* 29, 760, 1995.

5.   Weber, J. and May, P., *Am. J. Hum. Genet.,* 44, 388, 1989.

6.   Hamada, H., Seidman, M., Howard, B., and Gorman, C., *Mol. Cell. Biol.,* 4, 2622, 1984.

7. Gaillard, C. and Strauss, F., *Science,* 264, 433, 1994.

8. Pardue, M., Lowenhaupt, K., Rich, A., and Nordheim, A., *EMBO J.,* 6, 1781, 1987.

9. Weber, J., *Genomics,* 7, 524, 1990.

10. Hudson, T., Engelstein, M., Lee, M., Ho, E., Rubenfield, M., Adams, C., Housman, D., and Dracopoli, N., *Genomics,* 13, 622, 1992.

11. Economou, E., Bergen, A., Warren, A., and Antonarakis, S., *Proc. Natl. Acad. Sci. USA,* 87, 2951, 1990.

12. Gastier, J., Pulido, J., Sunden, S., Brody, T., Buetow, K., Murray, J., Weber, J., Hudson, T., Sheffield, V., and Duyk, G., *Hum. Mol. Genet.,* 4, 1829, 1995.

13. The Utah Marker Development Group, *Am. J. Hum. Genet.,* 57, 619, 1995.

14. Sheffield, V., Weber, J., Buetow, K., Murray, J., Even, D., Wiles, K., Gastier, J., Pulido, J., Yandava, C., and Sunden, S., *Hum. Mol. Genet.,* 4, 1837, 1995.

15. Kwiatkowski, D. and Diaz, M., *Hum. Mol. Genet.,* 1, 658, 1992.

16. Henske, E., Ozelius, L. Gusella, J., Haines, J., and Kwiatkowski, D., *Genomics,* 17, 587, 1993.

17. Ostrander, E., Jong, P., Rine, J., and Duyk, G., *Proc. Natl. Acad. Sci. USA,* 89, 3419, 1992.

18. Armour, J., Neumann, R., Gobert, S., and Jeffreys, A., *Hum. Mol. Genet.,* 3, 599, 1994.

19. Allen, R. C., Zoghbi, H. Y., Moseley, A. B., Rosenblatt, H. M., and Belmont, J. W., *Am. J. Hum. Genet.,* 51, 1229, 1992.

20. Busque, L. and Gilliland, D. G., *Blood,* 82, 337, 1993.

21. Willman, C. L., Busque, L., Griffith, B. B., Favara, B. E., McClain, K. L., Duncan, M. H., and Gilliland, D. G., *New Engl. J. Med.,* 331, 154, 1994.

22. Gale, R. E., Wheadon, H., and Linch, D. C., *Fr. J. Haematol.,* 79, 193, 1991.

23. Mashal, R., Lester, S., and Sklar, J., *Cancer Res.,* 53, 4676, 1993.

24. Knudson, A. G. J., *Proc. Natl. Acad. Sci. USA,* 68, 820, 1971.

25. Seizinger, B., Rouleau, G., Ozelius, L., Lane, A., St. George-Hyslop, P., Huson, S., Gusella, J., and Martuza, R., *Science,* 236, 317, 1987.

26. Mao, L., Lee, J., Fan, Y., Ro, J., Batsakis, J., Lippman, S., Hittelman, W., and Hong, W., *Nature Med.,* 2, 682,1996.

27. Zahn, L. and Kwiatkowski, D., *Genomics,* 28, 140, 1995.

28. Weber, J. and Wong, C., *Hum. Mol. Genet.,* 2, 1123, 1993.

29. Ausubel, F., Brent, R., Kingston, R., Moore, D., Seidman, J., Smith, J., and Struhl, K., *Curr. Protocols Mol. Biol.,* 2, 1994.

30. Modrich, P., *Science,* 266, 1959, 1994.

31. Han H. -J., Yanagisawa, A., Kato, Y., Park, J.-G., and Nakamura, Y., *Cancer Res.,* 53, 5087, 1993.

32. Risinger, J., Berchuck, A., Kohler, M., Watson, P., Lynch, H., and Boyd, J., *Cancer Res.,* 53, 5100, 1993.

33. Nawroz, H., Koch, W., Anker, P., Stroun, M., and Sidransky, D., *Nature Med.,* 2, 1035, 1996.

34. Chen, X., Stroun, M., Magnenat, J.-L., Nicod, L., Kurt, A.-M., Lyautey, J., Lederrey, C., and Anker, P., *Nature Med.,* 2, 1033, 1996.

35. Boland, C., *Nature Med.,* 2, 972, 1996.

36. Shibata, D., Navidi, W., Salovaara, R., Li, Z.-H., and Aaltonen L., *Nature Med.,* 2, 676, 1996.

37. Henske, E., Neumann, H., Scheithauer, B., Herbst, E., Short, M., and Kwiatkowski, D., *Genes Chrom. Cancer,* 13, 295, 1995.

38. Henske, E., Scheithauer, B., Short, M., Wollmann, R., Nahmias, J., Hornigold, N., Slegtenhorst, M. V., Welsh, C., and Kwiatkowski, D., *Am. J. Hum. Genet.,* 59, 400, 1996.

Chapter 2

Molecular Detection of Dynamic Mutations

R. Frank Kooy, Ben A. Oostra,
and Patrick J. Willems

Contents

I.   Introduction

Dynamic mutations are a new class of mutations characterized by an unstably inherited trinucleotide repeat. The mutations found in classical Mendelian diseases include stably transmitted point mutations, deletions, insertions, or duplications. In the new group of dynamic diseases, an unstable and amplified trinucleotide repeat is responsible for the phenotype.

Fragile X syndrome and spinobulbar muscular atrophy (SBMA) were the first disorders in which such a dynamic mutation was recognized.[1,2] In fragile X syndrome, there exists an expanded CGG repeat in the 5' untranslated region of the *FMR1* gene (*FMR1* for **fra**gile X **m**ental **r**etardation **1**). This repeat is highly polymorphic in the human population, with 6 to 53 repeat units. In fragile X patients, repeat numbers of 200 up to several thousand repeats have been found. Soon after this discovery in 1991, it became clear that the fragile X syndrome was only the first example of a new class of mutations, and the term "dynamic mutation" was introduced.[3]

Over the last 5 years, 11 dynamic mutations have been discovered (Table 2.1). Expanded trinucleotide repeats are responsible for clinical abnormalities and/or cytogenetic expression of fragile sites. Of these dynamic mutations, 9 are associated with 11 distinct disorders. The dynamic mutations can be classified into three groups (Table 2.1). The first group consists of the fragile sites FRAXA (**fra**gile site, **X** chromosome, **A** site), FRAXE (**fra**gile site, **X** chromosome, **E** site), FRAXF (**fra**gile site, **X** chromosome, **F** site), FRA11B (**fra**gile site, chromosome **11**, **B** site), and FRA16A (**fra**gile site, chromosome **16**, **A**

TABLE 2.1
Triplet Repeat Mutations

Condition	Chromo- somal Location	Repeat Sequence	Repeat Locali- zation	Normal	Alleles Premu- tation	Full Mutation
FRAXA	Xq27.3	CCG	5' untranslated	6–53	43–230	>200
FRAXE	Xq28	CGG	?	6–35		>200
FRAXF	Xq28	CGG	?	6–29		300–500
FRA11B	11q23.3	CGG	5' untranslated	8–14	80–100	>100
FRA16A	16p13.11	CGG	?	16–49		1000–2000
SBMA	Xq11–12	CAG	ORF	11–33		36–66
Huntington	4p16.3	CAG	ORF	9–34	30–35	36–121
SCA1	6p22–23	CAG	ORF	6–39		40–81
DRPLA/HRS	12pter–p12	CAG	ORF	3–28		49–75
Machado-Joseph/ SCA3	14q32.1	CAG	ORF	7–40		61–84
Myotonic dystrophy	19q13.3	CTG	3' untranslated	3–37		35–2000

site). These fragile sites are all associated with large expansions (>200 repeats) of a CGG repeat located on a CpG island that becomes hypermethylated following repeat expansion. In fragile X syndrome, hypermethylation downregulates the expression of FMR1, leading to severely diminished amounts of fragile X protein (FMRP). This loss-of-function mutation is responsible for the clinical abnormalities in fragile X patients, including mental retardation. A number of families in which FRAXE segregates with a mild form of mental retardation have been reported.[4] Chromosome breaks near FRA11B result in a partial loss of the long arm of chromosome 11 and cause Jacobsen syndrome, which is characterized by mental retardation and dysmorphic features.[5] The folate-sensitive fragile sites FRAXF[6,7] and FRA16A[8] have not been found to be associated with a clinical phenotype. A second group consists of seven neurodegenerative disorders, including SBMA,[2] Huntington's disease (HD),[9] spinocerebellar ataxia 1 (SCA1),[10] Machado-Joseph disease (MJD)[11] which is allelic to spinocerebellar ataxia 3 (SCA3),[12] and the allelic disorders dentatorubral-pallidoluysian atrophy (DRPLA)[13,14] and Haw River syndrome (HRS).[15] In this group, the disease is caused by an expanded CAG repeat expansion in the open reading frame of the respective genes. This results in mutant proteins with large polyglutamine stretches associated with cell death in specific neurological tissues, presumably as a consequence of a gain-of-function of the mutant proteins. Apart from the CAG repeat, these five genes share little sequence homology. All CAG-associated dynamic mutations are generally of late onset and have an autosomal dominant mode of inheritance, with the exception of SBMA, which has an X-linked recessive mode of inheritance. Compared to CGG repeat expansion, CAG repeat expansion is limited with repeats rarely exceeding 100 units. The clinical spectra of some of these disorders overlap, hampering differential diagnosis based upon clinical criteria.

Myotonic dystrophy, the most common autosomal-dominant adult muscular dystrophy, cannot be classified in either of these two groups. A CTG repeat is present in the 3′ untranslated region of the DM gene, which is a protein kinase.[16] Expansion of this CTG repeat can be dramatic, and repeat numbers of more than 8000 have been reported. The expanded CTG repeat is not hypermethylated and does not lead to the expression of a fragile site.

Apart from these differences, the dynamic mutations share many characteristics. Dynamic diseases are rarely caused by a new mutation but are nearly always inherited from one of the parents. Trinucleotide repeats in a the normal size range are transmitted without size alteration. Amplified repeats in the premutation and certainly in the full mutation range are meiotically unstable when transmitted through generations. Premutated alleles have a size between that of normal and mutated alleles and are not associated with expression of the disease and/or fragile site, but they are prone to expansion (or sometimes contraction) when transmitted to the next generation. Premutations have only been described for fragile X syndrome, Jacobsen syndrome, and a few families with Huntington's disease.[5,17-19] For CAG repeat disorders, as well as for myotonic dystrophy, an inverse correlation is found between repeat size, on one hand, and severity and age of onset of the disorder, on the other hand. The largest repeats are found in congenital cases, the smallest in patients with a late age of onset. The allele size by itself, however, cannot be used as an individual diagnostic criterion, as the spread of the age of onset in patients with specific allele size is large.[20-22]

The consequence of repeat amplification when transmitted from generation to generation is a phenomenon called anticipation. Anticipation is observed in families affected by any dynamic disorder. In the case of CAG and CTG repeat disease, the younger generations have larger repeats and are in general more severely affected at an earlier age of onset.[20-22] As a result of an increase in CGG repeat length, younger generations of fragile X families have a higher chance of being affected.[19,23,24]

Interestingly, the rate of repeat expansion is different in maternal and paternal transmissions. In Huntington's disease, for instance, paternal transmission is associated with an average 20-fold greater CAG repeat size increase than maternal transmission;[20] whereas, in fragile X syndrome, the repeat only increases to large repeat sizes during maternal transmission and remains stable during paternal transmission in the premutation range.[25,26] Mutations other than trinucleotide repeat expansion have not yet been found in any of the dynamic diseases with the exception of fragile X syndrome, which can also be caused by deletions or mis-sense mutation (Figure 2.1).

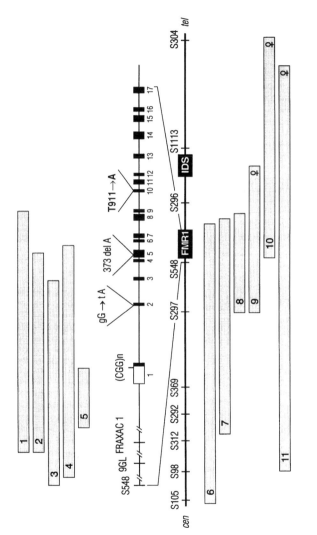

FIGURE 2.1

Mutations causing fragile X syndrome. See Reference 95 for gross deletion patient 1, Reference 96 for patient 2, Reference 97 for patient 3, Reference 98 for patient 4, Reference 99 for patient 5, Reference 100 for patient 6, Reference 101 for patient 7, Reference 102 for patient 8, Reference 103 for patient 9, Reference 104 for patient 10, and Reference 105 for patient 11. Other mutations include a gG-to-tA nucleotide substitution destroying a splice site,[40] a frameshift mutation due to deletion of an adenosine reside at position 373 in exon 5,[40] and a T→A point mutation substituting a highly conserved isoleucine for an asparagine in exon 10.[39] Females are indicated by the ♀ sign.

II.   Dynamic Mutations

A.   CGG Repeats

1.   Fragile X syndrome

Fragile X is the most common form of inherited mental retardation, with an estimated frequency of 1/4000 in males.[27] The name is derived from the presence of a fragile site at chromosome Xq27.3 which becomes visible after cultivation of lymphocytes from patients in folate-deprived media. The fragile X phenotype can be characterized by macroorchidism, long face, prominent forehead, large ears, and moderate to severe mental retardation. However, not all patients display such a characteristic phenotype, and differential diagnosis with other forms of (X-linked) mental retardation is often not feasible on the basis of the phenotype alone, certainly not in females. The pattern of inheritance is best described as X-linked dominant with reduced penetrance in females, as only 60 to 70% of the females with an amplified repeat in the full mutation range are affected, albeit often less severely than their male relatives.[28]

We now know that many peculiar features of fragile X inheritance are due to an unstable CGG repeat in the 5' untranslated region of the fragile X gene, *FMR1*.[1,19,29-32] The repeat is stably transmitted and polymorphic in the normal population, with allele sizes ranging from 6 to 53 repeats. Some unaffected individuals have larger and unstable repeats that expand when transmitted from parent to offspring (repeat regression is less often observed). These repeats are called premutations, which can be as small as 43 repeats[33] (smaller than a large stable allele) and as large as 230 repeats[34] (larger than small full mutations). Individuals with a premutation have normal levels of *FMR1* mRNA and *FMR1* protein (FMRP) and consequently are not affected.[35-37] Premutated repeats can expand into the full mutation range of 200 to 3000 repeats. These full mutations are associated with hypermethylation of a CpG island 5' of the untranslated region of the *FMR1* gene containing the CGG repeat. As a result, no mRNA is transcribed, and hence no *FMR1* protein is present. Amplification of the CGG repeat in fragile X syndrome is, therefore, a loss-of-function mutation. In some unaffected individuals, large, unmethylated premutations exist that have more repeats than small, methylated full mutations.[34,38] The premutations are unstably transmitted and increase in size in most intergenerational transmissions. Interestingly, the repeat size only increases to a full mutation when transmitted from mother to child and never when transmitted from the father. Males carrying a premutation are therefore called normal transmitting males (NTMs). Their daughters are never affected, although their grandchildren are at risk.[23,24] The finding that the full mutation found in all tissues from affected males is not present in their sperm explains why daughters of affected males are carriers of a premutation only.[25] The chance that a premutation expands to a full mutation when transmitted through a female depends on its size: repeats larger than 90 triplets have a >99% chance of expanding to a full mutation in the next generation; premutations smaller than 60 triplets have only a small risk (<1%).[19]

As fragile X syndrome is common among mentally retarded patients, and the phenotypical characteristics can be vague, many laboratories nowadays routinely screen all patients with mental retardation for the presence of an enlarged CGG repeat in *FMR1*. This identifies the vast majority (>99%) of fragile X patients.

A few patients have been described without CGG repeat enlargement. Molecular analysis revealed deletions of the whole or part of the *FMR1* gene. All deletions described until now are summarized in Figure 2.1. Three intragenic mutations have been identified: a patient with a mis-sense mutation,[39] a patient with a frameshift mutation, and a patient with a double nucleotide substitution destroying a splice site.[40] The patient with the mis-sense mutation is peculiar, as, although mutated *FMR1* protein is present,[41] his phenotype is particularly severe, suggesting this mutant protein may have a deleterious gain-of-function mutation.[39,42]

2. FRAXE syndrome

The FRAXE fragile site at Xq28 was recognized in 1992 as being distinct from the FRAXA site at Xq27.3.[43] The two sites cannot be discriminated by routine cytogenetic analysis due to their proximity (600 kb apart). Most individuals diagnosed with FRAXE were originally falsely diagnosed as being fragile X patients on the basis of cytogenetic expression of a fragile site at Xq27.3-Xq28. Absence of *FMR1* trinucleotide expansion led to the suspicion of FRAXE syndrome, which could be confirmed after the discovery of the dynamic mutation leading to FRAXE. Like FRAXA expression, FRAXE expression is associated with amplification of a CGG repeat embedded in a CpG island.[4] Normal individuals have 6 to 35 copies of this repeat. Males with more than 200 copies of the repeat show cytogenetic expression of the site and methylation of the CpG island. Females with a similar repeat amplification do not always show cytogenetic expression of FRAXE. Little is known about the clinical phenotype of FRAXE-expressing individuals. No clinical phenotype was recognized in the original description of FRAXE,[43] but mild mental retardation segregating with FRAXE expression was reported 2 years later in an extended study of the same family.[44] The frequency of FRAXE syndrome is unknown, and only 11 families in which FRAXE segregates with mild mental retardation have been described.[4,44-46] The developmental delay in females with a large repeat is generally mild or even absent. Apart from the mental retardation, no other clinical symptoms are shared between affected persons, and it is not clear if FRAXE syndrome is really associated with a clinical phenotype. The FRAXE repeat may contract when transmitted paternally and contract or increase when transmitted maternally.[46] No gene near the CGG repeat and the CpG island has been identified yet; however, two patients with overlapping submicroscopic deletions near FRAXE have been described, suggesting the possible presence of a gene near the CGG repeat of FRAXE.[47]

3. FRAXF fragile site

A third fragile site, just distal from FRAXE and within 3 Mb from FRAXA, was described in 1993.[6] This fragile site was also found to be the consequence of CGG repeat expansion. Like the CGG repeat at the FRAXA and FRAXE sites, the repeat is part of a CpG island that becomes methylated after repeat expansion.[7] Control persons have repeat sizes of 6 to 29 repeats, while persons showing cytogenetic expression of the site have repeat sizes of 300 to 500 repeats. FRAXF appears not to be associated with any clinical abnormalities. Eventual instability upon transmission remains to be demonstrated, as very few families with this presumably rare fragile site have been described. A FRAXF site should be suspected when a patient shows cytogenetic expression of a fragile site at Xq27.3-Xq28, and FRAXA and FRAXE have been excluded.

4. FRA11B fragile site

Jacobsen syndrome is a rare chromosome deletion (11q–) syndrome. It is characterized by severe mental retardation and specific dysmorphic features.[49] The break has been shown recently to occur near the rare fragile site FRA11B, caused by CGG repeat expansion within the protooncogene *CBL2*. As both FRA11B and Jacobsen syndrome are extremely rare, it is unlikely that they occur in a single family by chance. Therefore, expansion of the CGG repeat is likely to be involved in the chromosome breakage observed in Jacobsen syndrome patients.[5] Like other fragile sites, the FRA11B CGG repeat is close to a CpG island and is methylated once the repeat exceeds a certain threshold length.

5. FRA16A fragile site

This folate-sensitive fragile site is rare in Caucasians and has not been observed in other populations.[8,48] No phenotypic abnormalities have been found in individuals expressing this site. FRAX16A is of considerable interest, as elucidation of the molecular nature of FRA16A demonstrated a basis similar to the fragile sites located on the X chromosome, namely a CGG repeat in a CpG island. This repeat is polymorphic but small in the normal nonexpressing population, whereas in FRA16A-expressing individuals, this repeat has expanded. Like the fragile sites on the X chromosome, the repeat becomes methylated upon expansion. The latter finding is interesting, as this region of chromosome 16 is not a site of methylation in nonexpressing individuals, in contrast to the CpG islands near the fragile sites on the X chromosome, which also can be methylated in normal females during Lyonization.

B. CAG Repeats

1. Spinal and bulbar muscular atrophy

X-linked spinal and bulbar muscular atrophy, or Kennedy's disease, is an X-linked recessive disease of motor neurons. Typical characteristics of SBMA include fasciculations followed by proximal muscle weakness and atrophy, bulbar signs, intentional tremor, and dysphagia. In general, the disease is of late onset, is slowly progressive, but does not reduce life expectancy.[50] Males quite often also show signs of gynecomastia and reduced fertility.[51] The latter symptoms led La Spada et al.[2] to investigate whether the androgen receptor gene (AR), which is located in the same chromosomal region as SBMA (Xq11-12), might be involved in the disease. The androgen receptor is part of a supergene family of steroid-binding intracellular receptors that bind to specific sites of DNA in the nucleus, thereby regulating transcription of specific genes. Amplification of a polymorphic CAG repeat was found in all patients with SBMA. The CAG repeat is located within the coding region of the AR gene, where it encodes a polyglutamine stretch. Two groups of repeat sizes have been reported: normal individuals with 11 to 33 repeats, and SBMA patients with 36 to 66 repeats. Individuals with intermediate-sized alleles have not yet been identified. There is a correlation between repeat length and severity of the disease, but other genetic or environmental factors may also contribute to the variability in clinical manifestations.[52,53] The amplified repeat is transmitted unstably; both expansions and contractions are observed, the greater instability being observed in paternal transmissions.[54,55]

Although the clinical phenotype of SBMA is rather specific, the diagnosis may be confirmed by DNA analysis, and presymptomatic carriers might be identified and prenatal diagnosis offered. As three cases of SBMA have been identified in a group of 25 individuals with unclassified motor neuron disease,[56] it might be worthwhile to screen patients with less specific neurologic symptoms as well by determination of the CAG repeat size in the AR gene. CAG repeat expansion is the only mutation described causing SBMA. Deletions or other mutations in the AR gene cause different syndromes, such as testicular feminization syndrome or micropenis with gynecomastia.

2. Huntington's disease

Linkage of the autosomal-dominant Huntington's disease to the chromosomal region of 4p16.3 was reported in 1983.[57] Patients suffer from motor disturbance, cognitive loss, and psychiatric manifestations. This rare disease typically has its onset in midlife and is progressive and inevitably fatal.[58] The Huntington's collaborative research group reported the identification of the HD gene in 1993. The gene has a polymorphic CAG repeat in the 5' translated region, encoding a polyglutamine stretch.[9] Patients have a repeat size above a threshold of 36, whereas repeat sizes of 9 to 34 are found in healthy individuals. A factor complicating repeat size determination is the existence of a CCG repeat 3' to the CAG repeat. The primers used to amplify the CAG repeat also amplify this CCG repeat, which was originally thought to be of a fixed size but was more recently found to have at least eight alleles of different sizes. As the length of this CCG repeat has no influence on the phenotype of the patient, the length of this CCG repeat should be determined in individuals with borderline repeat sizes.[59]

An inverse correlation between age of onset and repeat length is well established.[20,61,62] In nearly all patients suspected of Huntington's disease on the basis of their clinical appearance, an enlarged CAG repeat has been identified. A few families are described with clinical manifestations, but without CAG repeat expansion.

It remains unclear, however, whether these patients represent phenocopies of HD, as mutations other than CAG repeat expansion in huntingtin have not been reported. A few patients with amplified CAG repeats in the huntingtin gene on both chromosomes have been reported, but this has no effect on the severity of the phenotype.[63] Deletion of one huntingtin allele does not result in HD or any other abnormality, indicating that the amplified CAG repeat has a gain-of-function which is deleterious for certain brain cells.[9]

The CAG repeat of Huntington's patients inherits unstably in over 85% of transmissions. No changes in length in normal chromosomes have been observed, although one would expect these to occur occasionally to explain the spread of allele sizes. New disease alleles occasionally arise from transmissions of intermediate-sized alleles with repeat lengths of 30 to 35 repeats.[17,18,64] The individuals carrying intermediate alleles themselves are asymptomatic, but their children are at risk of having the disease. Maternal transmissions are associated with size alternations of 5 repeat units or less, size increase being more common than size decrease. Paternal transmission is predominantly associated with size increase of on average 10 repeat units, but decreases comparable in size to those found in maternal transmissions also occur.

3.   Spinocerebellar ataxia type 1

SCA1 is also an autosomal-dominant neurodegenerative disorder caused by expansion of a CAG repeat within the coding region of a gene called *ataxin-1*. This gene is located in the chromosomal region 6p22-23.[10,65] The main clinical symptoms of SCA1 include cerebellar ataxia and upper motor neuron signs. Some of these symptoms are shared between other forms of spinocerebellar ataxias or other CAG repeat disorders;[66] therefore, the detection of an enlarged CAG repeat in the coding region of the ataxin-1 gene can be used to confirm the diagnosis of SCA1. Normal individuals have between 6 and 39 CAG repeats, whereas patients have 40 to 81 repeats. No overlap between these two groups has been identified so far. As in HD, the repeat length is inversely correlated with age of onset but in itself is an insufficient parameter to be of prognostic value, as the range is large and factors other than repeat length may be involved in the severity of the disease. As in HD, the repeat is inherited unstably in transmissions through both sexes. Approximately two thirds of paternal transmissions are accompanied by an alteration in the repeat size. Increases in size of up to 28 units have been reported, but decreases of a few repeat units occasionally occur. Maternal transmissions are stable in two thirds of all reported cases, and if size alterations occur these are on average smaller (up to 6 repeat units), and shortening is more often observed than elongation.[67,68]

4.   Dentatorubral-pallidoluysian atrophy

DRPLA is an autosomal dominant syndrome with myoclonus epilepsy, dementia, ataxia, and choreoathetosis.[69] An expanded CAG repeat within the coding region of cDNA clone CTG-B37 at 12pter-p12 was found to be responsible for this disorder. Although few affected families have been analyzed yet, an inverse relation between repeat length and age of onset and instability of the repeat upon paternal and maternal transmission is well established. In general, paternal transmissions are associated with size increase and maternal transmissions with size decrease.[13,14,70]

Originally, it was thought that DRPLA was restricted to Japan, where it is estimated to be responsible for about 2.5% of cases with spinocerebellar abnormalities;[71] however, recently DRPLA was also diagnosed in three Danish patients. These patients had initially been diagnosed as HD patients, but no expanded CAG repeat in the HD gene was present.[72] Dentatorubral-pallidoluysian atrophy was also known as Haw River syndrome in the U.S. (see next paragraph).

5.   Haw River syndrome

Haw River syndrome is now known to be caused by expansion of the same CAG repeat that causes DRPLA.[15] It was first described in a family living in a town called Haw River in North Carolina. Clinical abnormalities similar to those found in DRPLA were present in that family.[73] Another HRS family of Afro-American origin was found without myoclonus and seizures, demonstrating the variety of clinical symptoms that may occur in different families.

6.   Machado-Joseph disease

Machado-Joseph disease is another dominantly inherited ataxia, accompanied by fasciculations, dystonia, and tremors. The clinical spectrum overlaps with the other ataxias, and in part with other CAG repeat disorders.[74] Linkage to the chromosomal region 14q32.1 proved that MJD is different from SCA1 and SCA2 located on chromosomes 6 and 12, respectively.[75] The MJD gene contains a CAG repeat with significantly longer CAG alleles in patients as compared to healthy individuals (Table 2.1). Patients with longer repeats have an earlier age of onset than patients with shorter alleles. As is true for other CAG repeat disorders, paternal transmission of disease alleles is associated with a larger size increase than is maternal transmission.[11]

7.   Spinocerebellar ataxia type 3

Although spinocerebellar ataxia type 3 was thought to be clinically distinct from MJD, the localization of both disorders to the same region on chromosome 14 suggested the two disorders might be allelic.[76] Size increase of the MJD CAG repeat was found in large SCA3 kindreds from Germany and The Netherlands.[12,77] This demonstrates once more the variety of symptoms that may arise as the result of a single mutation.

C.   CTG Repeats

1.   Myotonic dystrophy

Myotonic dystrophy, or Steinert disease, is one of the most common muscular dystrophies, with an incidence of 1 in 8000. This autosomal-dominant disorder with variable penetrance is characterized by myotonia, muscle wasting, cataract, and hypogonadism. The age of onset is variable but in most cases occurs in mid life. Congenital cases, mostly observed after maternal transmission of the disease allele, are associated with a more severe clinical picture, including mental retardation and sometimes lethality.[78] The discovery of an enlarged CTG repeat in the 3' untranslated region of the DM-protein kinase gene (DM-PK) in patients has improved our understanding of the inheritance pattern observed in families. The repeat is highly polymorphic in the normal population, with allele sizes of 3 to 37 repeat units. In patients, repeats of from 35 to several thousand triplets are observed. The severity of the disease and the age of onset are correlated with the CTG repeat size. The increase in repeat length, associated with an increased expression of the disease (anticipation) in multigenerational MD families is more pronounced than for the other triplet disorders.[16]

III.   Molecular Diagnosis

A.   Detection of CGG Repeats

1.   Cytogenetic detection

Traditionally, the diagnosis of fragile site-associated syndromes, predominately fragile X syndrome, relied on the identification of a fragile site detectable in chromosome spreads of lymphocytes, fibroblasts, or amniocytes cultured in folate-deprived media. Molecular detection has largely replaced cytogenetic analysis, because (1) only some females with a full mutation show cytogenetic expression of the fragile site, (2) premutations are not associated with a fragile site and therefore are not detectable by cytogenetic analysis, (3) prenatal diagnosis based upon cytogenetic analysis of chorion villi or amniocytes is unreliable, and (4) cytogenetically, FRAXA cannot be discriminated from FRAXE or FRAXF.

2.   Polymerase chain reaction (PCR) detection

PCR-based determination of the length of a CGG repeat associated with a fragile site is a rapid way to detect amplified alleles in FRAXA, FRAXE, FRAXF, FRA11B, and FRA16A. Although amplification of a CGG repeat up to several kilobases in length has been described,[79] most laboratories have significant difficulties in amplifying CGG stretches of more than 150 repeats because of the CG-rich nature of the repeat. PCR amplification of smaller CGG repeats is feasible in 15-μl reaction volumes containing 3 pmol of each flanking primer (see Table 2.2); 10 mM Tris-HCl, pH 9.0; 50 mM KCl; 2.5 mM MgCl$_2$; 0.1% Triton X-100; 0.01% gelatin; 10% DMSO; 0.2 mM each of dATP, dCTP, and dTTP; 0.4 mM 7'-deaza-dGTP; 4 μCi [alpha-^{32}P]dCTP; 0.8 unit Taq polymerase; and 50 to 100 ng template DNA of good quality. The mixture is topped with a drop of mineral oil. The Taq polymerase has to be added to this mix after an initial denaturation of 10 min at 95°C, as the high denaturation temperature may inactivate the enzyme before amplification if added to the reaction mix immediately. Thermal cycling may be carried out for 33 cycles and consists of denaturation at 95°C for 1 min, annealing at 65°C for 1.5 min, and extension at 72°C for 2 min. The PCR reaction is followed by a final extension step at 72°C for 5 min. For the amplification of CGG repeats, a denaturation temperature of 95°C is essential. With a lower denaturation temperature, amplification is unsuccessful, whereas a higher denaturation temperature destroys the Taq polymerase. In practice, not all PCR machines reach this temperature during the PCR program, and as a consequence the CGG alleles will not amplify. The use of thin-walled reaction tubes that easily transmit temperature changes is recommended. After PCR amplification, 3 µl of loading buffer (95% formamide, 0.09% bromophenol blue, 0.09% xylene cyanol FF) are added to 3.5 µl of sample. This mixture is incubated at 95°C for 5 min and loaded on a 6% denaturing polyacrylamide gel. Electrophoresis is stopped when the xylene cyanol FF dye has passed through about 3/4 of the gel. The gel is then removed and dried, covered with plastic foil, and exposed to an X-ray sensitive film at –80°C for 24 to 72 hr.

Interpretation of the X-ray films is not straightforward, as very large alleles that may be present will not amplify. After amplification of the CGG repeats on the X chromosome, males show a normal-sized allele, a premutation, or no amplification. A normal-sized allele indicates the absence of repeat expansion. Premutations in fragile X syndrome can also be detected, but a premutation being found might indicate mosaicism for a premutation and a full mutation which is present in many males. In case of absent amplification, it is likely that a full mutation, too large to amplify, is present; however, the sample may simply have failed to amplify for technical reasons, or a deletion near the CGG repeat might have prevented primer attachment. As females have two X chromosomes, their genotype is unclear when a single band of normal size is visible. In that case, it is not possible to discriminate among homozygosity for a normal allele, heterozygosity for a normal allele and a full mutation allele too large to be amplified, or hemizygosity for a normal allele and a deletion of the CGG repeat region. Southern blots are used to detect full mutations and hemizygosity.

PCR detection is, therefore, a rapid way to screen large numbers of samples and accurately determine the size of a normal or premutated allele, but identification of the full mutation in patients is not possible. Therefore, the use of Southern blot is still advocated in the diagnosis of fragile X syndrome and other CGG repeat disorders.

3.   Southern blot detection of fragile X syndrome

Southern blot analysis allows the detection of normal, premutation, or full mutation alleles; however, discrimination between large alleles in the normal size range and small premutations and the estimation of exact repeat sizes are not possible using this technique. Southern blot of DNA digested with methylation-sensitive enzymes is also used to detect the methylation status of the CpG island containing the CGG repeat. Determining the methylation status is important to discriminate large premutations from the smallest full mutations. The distinction between these two alleles cannot be made on the basis of the CGG repeat size alone, because the sizes of large premutations and small full mutations overlap. Determination of the methylation

TABLE 2.2
Primer Pairs for Amplification of Trinucleotide Repeats

Condition	Primer Name	Primer Sequence	Annealing Temperature (°C)	Size PCR Product Without Repeat	Ref.
FRAXA	c: f:	GCTCAGCTCCGTTTCGGTTTCACTTCCGGT AGCCCCGCACTTCCACCACCAGCTCCTCCA	65	221	19
FRAXE	598: 603:	GCGAGGAAGCGGCGGCAGTGGCACTGGG CCTGTGAGTGTGTAAGTGTGTGATGCTGCCG	65	288	4
FRAXF		CAGCCTGCGCCTTACAGCGGGTTCATGGCG AGGGGCAGGGGCGGTGGCTCAGGTTTCTC	65	133	7
FRA11B		CTCCCTCCGCCGGATAGCC TGAAGGAGGGGCCTCTCCCG	65	40	5
FRA16A	a: b:	GCCGGCTGCCGCTCGGGCTCCCGCT CGGGTCCCTGCCCGTCTGAAAA	65	114	8
SBMA		TCCAGAATCTGTTCCAGAGCGTGC GCTGTGAAGGTTGCTGTTCCTCAT	60	222	2
Huntington	Hu3: Hu4:	GGCGGCTGAGGAAGCTGAGGA ATGGCGACCCTGGAAAAGCTGATGAA	62	107[a]	91[b]
SCA1	Rep-1: Rep-2:	AACTGGAAATGTGGACGTAC CAACATGGCAGTCTGAG	66	125	10
DRPLA/HRS		CACCAGTCTCAACACATCACCATC CCTCCAGTGGGTGGGGAAATGCTC	62	96	92

Machado-Joseph/SCA3	MJD52:	CCAGTGACTACTTTGATTCG	62	199	11
	MJD70:	CTTACCTAGATCACTCCCAA			
Myotonic dystrophy	409:	GAAGGGTCCTTGTAGCCGGGAA	68 ·	49	93
	410:	AGAAAGAAATGGTCTGTGATCCC			

[a] This includes a CCG repeat immediately downstream of the CAG repeat. This CCG repeat is polymorphic, consisting of 5 to 12 repeats; 107 is the size of the PCR product including the $(CGG)_7$ of the most frequent allele. This allele is in strong linkage disequilibrium with HD chromosomes. The number of CCG repeats can be determined using primers HD4F: GCAGCAGCAGCAGCAACAGCCGCCACCGCC and HD5R: CTTTCTTTGGTCGGTGCAGCGGCTCCTCAG. Amplification can be performed in the reaction mixture described for amplification of CGG repeats with 1-μg template DNA. Thermal cycling is performed, after a first denaturation step at 95°C for 5 min, for 38-40 cycles consisting of a denaturation step at 95°C for 1 min and a combined annealing/elongation step at 70°C for 2 min. The PCR product would be 161 bp if no CCG repeats were present.[94]

[b] The Hu4 primer reported did not match the sequence published for the HD gene; therefore, an extra A was added to match the sequence of the HD gene: ...GAAAAGC.... .

status of the CpG island can distinguish between unmethylated transcribed alleles and methylated alleles that are not transcribed and result in fragile X phenotype.

For optimal Southern blot analysis, two 8-µg samples of DNA are digested with the restriction endonucleases BglII and HindIII or HindIII/EagI, respectively. The samples are separated according to size on a 0.7% agarose gel in 1× TAE: (1× TAE: 0.04 M Tris-acetate, 1 mM EDTA) or 0.5× TBE (1× TBE: 0.9 M Tris base, 0.9 M boric acid, 2 mM EDTA) in combination with a size marker at 1 to 3 V/cm for 12 to 18 hr. The DNA is stained with ethidium bromide (final concentration: 0.5 µg/ml) after migration through the gel to enhance fragment resolution. For optimal transfer of the >10-kb fragments of the BglII digest, the gel is soaked in 0.25 M HCl for 20 min before blotting. After neutralization in three brief rinses of tap water, the DNA is transferred onto a membrane (Southern blotting) according to the instructions of the manufacturer and hybridized overnight in 0.1 to 0.2 ml/cm^2 hybridization mix (hybridization mix: 6× SSC, 5× Denhardt's, 0.1% SDS; 1× SSC: 0.15 M NaCl, 0.015 M Na$_3$Citrate [pH = 7.0]; 1× Denhardt's: 0.02% Ficoll, 0.02% bovine serum albumin, 0.02% polyvinylpyrrolidone) with radioactively labeled probe pP2 (or equivalent). pP2 is the same probe as StB12.3 or pfxa7. Other probes that can be used are pfxa3 (= Ox0.55 = PX6), Ox1.9, or pE5.1 (= pfxaI). All the probes hybridize to the region between the CGG repeat and the proximal translated region of the *FMR1* gene (for review, see Reference 80). After hybridization for 12 to 18 hours at 65°C, the membrane is washed sequentially for 15 min twice in 3× SSC, 0.1% SDS; twice in 1× SSC, 0.1% SDS; and twice in 0.3× SSC, 0.1% SDS at 65°C. SDS is removed by a brief rinse in 2× SSC at room temperature; the membrane is sealed in plastic foil and exposed to an X-ray-sensitive film with an intensifying screen at –80°C for 24 to 72 hr.

The HindIII digest generates a fragment of 5.2 kb in control persons. In case of a large difference in size of the two normal *FMR1* alleles (e.g., 12 and 40 repeats), a double band around 5.2 kb might be visible. Affected individuals have 200 or more repeats and hence HindIII fragments that are more than 5.7 kb. Due to the mitotic instability of the full mutation, the amplified repeats have large size differences (200 to 1000 repeats), visible as a smear on the film. This smear can be very diffuse and sometimes barely visible on HindIII blots. A BglII digest is therefore performed, as this generates a larger 12-kb fragment, compressing the smear to a distinct band which is easily visible (Figure 2.2A). It is recommended that a long exposure of each film be made, as extensive smearing occasionally occurs and these smears may be overlooked after shorter exposures; however, a BglII digestion alone is not sufficient for diagnostic purposes, as it is not able to detect small expansions. A HindIII/EagI double digest is essential when the number of repeats is approximately 200, because distinction between a large premutation and a small full mutation depends on the methylation of the CGG repeat. In control males, EagI will cut the unmethylated CpG island proximal to the *FMR1* gene, generating fragments of 2.8 and 2.4 kb (only the 2.8-kb fragment is visible) in males. Because of the methylation of one of the X chromosomes by the Lyonization process in females, half of the 5.2-kb fragments are not cut by EagI, leading to the presence of both a 5.2-kb

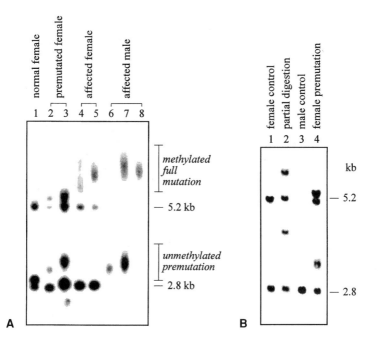

FIGURE 2.2

Molecular diagnosis of fragile X syndrome using Southern blot after *Eco*RI/*Eag*I double digestion. (**A**) Due to Lyonization, a methylated 5.2 *Eco*RI band and an unmethylated 2.8-kb double digested band are visible in female controls. Due to differences in CGG repeat size between the maternal and the paternal allele, the 2.8-kb band is a doublet (lane 1). Four bands are visible in female carriers of a premutation, as the size difference between a normal allele and a premutated allele allows clear size separation on the gel, and because premutations, like normal alleles, are not methylated. Due to Lyonization, approximately half of the normal and the premutated alleles are methylated (lanes 2 to 3). Affected females show only three bands. The normal allele has a methylated 5.2-kb fragment and an unmethylated 2.8-kb fragment. The full mutation allele is visible as a smear larger than 5.7 kb, which is fully methylated (lanes 4 to 5). Affected males show a methylated smear larger than 5.7 kb (lanes 7 to 8). Some affected males are mosaic for a premutation and a full mutation (lanes 6 to 7). The male patient in lane 6 illustrates the need for a long exposure of the X-ray film, as the faint smear of the full mutation is barely visible. (**B**) After *Eco*RI/*Eag*I double digestion, a single unmethylated 2.8-kb band is visible in normal male controls (lane 3). The female control shows an unmethylated and a methylated (due to Lyonization) fragment (lane 1). Four bands are visible in a female carrier of a premutation, as explained in A (lane 4). The control in lane 2 may be misdiagnosed as a carrier of a full mutation, but the extra 6.5- and 4.1-kb bands are caused by incomplete *Eco*RI digestion. Partial *Eco*RI digests are occasionally observed and can be avoided by the use of *Hind*III instead of *Eco*RI.

(methylated by the Lyonization) and a 2.8-kb (not methylated by the Lyonization) fragment in female controls. In male patients with a fully methylated enlarged repeat, *Eag*I will not cut the *Hind*III fragment, and no extra band will become visible on the gel. Rehybridization of the *Hind*III or *Bgl*II blot with a control probe such as pS8[81] is necessary should one want to identify hemizygous females. These female patients will have half the intensity of their pP2 band relative to the intensity of the pS8 band on the blots compared to females with two *FMR1* copies.

The use of a *Hind*III rather than the commonly used *Eco*RI digest[82] is recommended because we have noticed that one of the *Eco*RI sites is sometimes refractory to digestion, especially in the double digestion mixture with *Eag*I. This partial digested product may be confused with a full mutation and thus lead to a false positive diagnosis. This is illustrated in Figure 2.2B for an *Eco*RI/*Eag*I double digest hybridized with pP2. Such artifacts are avoided by the use of *Hind*III rather than *Eco*RI.

4.   Southern blot detection of FRAXE, FRAXF, FRA11B, and FRA16A

In principle, the same protocol can be used to detect amplification of the CGG repeats at the FRAXE, FRAXF, FRA11B, and FRA16A fragile sites. Amplification at the FRAXE site is detectable after digestion of genomic DNA with *Hind*III and hybridization with probe OxE20, generating a 5.2-kb fragment in controls. Methylation of the region can be monitored in a *Hind*III/*Eag*I double digest, cutting the 5.2-kb band to smaller fragments.[4] Deletions near FRAXE have been detected in two patients worldwide with probe pS8 (DXS296) on a *Eco*RI digest, the same probe that can be used as a control probe for fragile X deletion detection in females.[81] To detect FRAXF, digest with *Eco*RI to generate 5-kb fragments in normal individuals after hybridization with probe pR15.0. *Eag*I digestion reduces this fragment to a 1.5-kb and a 3.5-kb fragment at unmethylated sites.[7] It is possible to detect FRA11B by preparing a *Xba*I/*Bgl*II double digest in combination with the methylation-sensitive enzyme *Sac*II, blotting and hybridizing with probe XS.[5] Controls show a single band of 2.1 kb after double digestion and of 1.1 kb after triple digestion. The FRA16A site can be detected with probe pf16A3 after *Pst*I digestion, generating a 2.2-kb fragment in normal individuals. Methylation analysis in the original description of the FRA16A site was carried out by digestion with a *Pst*I/*Rsa*I/*Fnu*4HI triple digest, which was probed with a 650-bp *Nor*I-*Rsa*I fragment from pf16A3. This probe detects an 850-bp constant band in controls and several smaller fragments in fragile site-expressing individuals.[8]

5.   FMRP detection

Recently, a test to detect FMRP, the protein product of *FMR1*, in lymphocytes has been developed. This test enables direct identification of FMRP in blood. It is a quantitative test to score the presence or absence of FMRP in individual lymphocytes.[83] To stain the FMRP, blood smears are treated as follows:

1. Fixation in 3% paraformaldehyde, 10 min — dissolve 3 g paraformaldehyde in 100 ml 0.1 M phosphate buffer (Sörensen): 18 mM NaH$_2$PO$_4$, 77 mM Na$_2$HPO$_4$, pH 7.3; heat to 60–65°C to dissolve; and adjust pH to 7.3. Store at –20°C in the dark.

2. Permeabilization in 100% methanol, 20 min.

3. Wash in 0.1 M phosphate-buffered saline (0.1 M PBS: 150 mM NaCl, 10 mM Na$_2$HPO$_4$, 1.5 mM KH$_2$PO$_4$, pH = 7.3), containing 0.5% BSA and 0.02 M glycine (2 × 5 min).

4. Immunoincubation with mouse anti(*FMR1*) 1:250 in PBS, containing 0.5% BSA, 0.02 M glycine (60 min or longer).

5. Wash in 0.1 M PBS containing 0.5% BSA and 0.02 M glycine (3 × 5 min).

6. Incubation with goat anti-mouse IgG conjugated with biotin (DAKO, code E433), supplemented with 10% normal human plasma (60 min).

7. Wash in 0.1 M PBS, containing 0.5% BSA and 0.02 M glycine (3 × 5 min).

8. Incubation with streptavidin-conjugated alkaline phosphatase complex, prepared according to the instructions (DAKO, code K391) 1:100 in PBS, containing 0.5% BSA and 0.02 M glycine (45 min).

9. Wash in 0.1 M PBS, containing 0.5% BSA and 0.02 M glycine (4 × 5 min).

10. Wash in 0.1 M Tris-HCl, pH = 7.3 (5 min).

11 Incubation with substrate New Fuchsin (DAKO, code K596), with 0.1 mM levamisole (2 × 20 min), with a rinse in 0.1 M Tris-HCl, pH = 7.3, between the incubation steps.

12. Wash in tap water (2 min).

13. Counterstain with Gill's hematoxylin (2 sec).

14. Rinse in tap water (5 min).

15. Mount with aquamount.

All steps are performed at room temperature (18 to 24°C), except for longer (12 to 18 hr) incubation with anti(*FMR1*) at 4°C. Blood smears should be prepared within 6 hr after sampling, air dried, and, if not stained immediately, sealed and stored at –80°C. Several *FMR1* antibodies have been described, and a good signal-to-noise ratio is seen when using *FMR1* antibody 1a.[84] Once stained, the slides are investigated under a conventional light microscope with a blue filter. Only lymphocytes can be screened because of endogenous alkaline phosphatase activity in other white blood cells. Lymphocytes are recognized by their large, unlobbed nucleus, surrounded by a small ring of cytoplasm where the FMRP is located. In normal males, the cytoplasmic ring of nearly every lymphocyte shows red FMRP staining. In affected males with a full mutation, only a small minority of cells stain for FMRP (<10%), as most male patients have a low grade mosaicism for the nonstaining full mutation cells (majority) and the positively staining premutation cells (minority). The advantage of this test is that FMRP in individual cells, and hence the rate of mosaicism, can be determined. In males, no overlap in percentage of positive cells is seen between mosaic and normal males. In females, the risk of a false negative diagnosis exists, as the mutated X chromosome is often preferentially inactivated, resulting in a percentage of FMRP-stained lymphocytes greater than the expected

50%. The test also does not detect premutations, as these are normally transcribed and translated into FMRP.

6.   Prenatal diagnosis

In principle, Southern blot and PCR techniques can be used for prenatal screening. One should, however, be careful in interpreting the *FMR1* methylation status of chorion villi, as the CpG island associated with *FMR1* is not always methylated in the chorion villi, despite the presence of a full mutation in the fetus.[85] In that case, full mutations are not yet fully methylated at the time of prenatal diagnosis. This could lead to a false negative diagnosis when a small full mutation is classified as a large premutation, on the basis of absence of methylation in the chorion villi.

B.   Detection of CAG Repeats

1.   Polymerase chain reaction detection

Polymerase chain reaction detection is for the moment the only method for the molecular diagnosis of CAG repeat diseases. In contrast to CGG or CTG repeats that may reach sizes of several kilobases, amplification of CAG repeats has never been reported to exceed 121 repeats. This is a size that can normally be amplified by PCR under routine conditions, especially since CAG repeats are easier to amplify than CGG repeats. Southern blot detection is not feasible, as the differences in repeat size between patients and controls are too small to detect by this method. A protein detection test as described for fragile X syndrome is also not feasible, as no detectable difference in protein expression level between affected and unaffected persons has been found.

A number of PCR primers have been described to determine CAG repeat sizes for most disorders. Table 2.2 provides one primer pair for each disease. Amplification of any CAG repeat template can be achieved in 20-ml volumes in a reaction mixture containing 3 pmol of each primer; 10 mM Tris-HCl, pH 9.0; 50 mM KCl; 1.5 mM MgCl$_2$; 0.1% Triton X-100; 0.01% gelatin; 0.2 mM each dATP, dGTP, dCTP, and dTTP; 1 unit Taq polymerase; 100 to 300 ng template DNA. This should be topped with a drop of mineral oil. Thermal cycling may be carried out for 30 cycles after an initial denaturation step of 95°C for 5 min and consists of denaturation at 95°C for 1 min, annealing at a specific temperature (indicated in Table 2.2) for 1 min, and extension at 72°C for 1 min, followed by a final extension step at 72°C for 5 min. To visualize the PCR products, one of the primers may be end-labeled in a reaction mixture containing 100 pmol primer, 10 mM Tris-acetate, 10 mM MgAc, 50 mM KAc, 20 to 40 μCi [gamma-^{32}P]dATP, and 1 unit T4 kinase. Incubate this at 37°C for 30 min. Inactivate the enzyme by heat inactivation at 95°C for 5 min. Alternatively, radioactive label may be incorporated in the PCR product and not in the primer by adding 0.25 μCi [alpha-^{32}P]dCTP to the PCR mixture and lowering the unlabeled dCTP concentration to 0.02 mM.

Many laboratories prefer end-labeling over label incorporation, as only one of the two strands of the PCR product is labeled with end-labeling, in contrast to label incorporation where both strands are labeled. As the gel to visualize the PCR product is run under denaturing conditions, the DNA strands of the PCR product will separate. Because of the CAG repeat nature of the PCR product, both strands have a different mobility on the gel. A single PCR product appears thus as two separate strands on the autoradiograph. As with end-labeling, only one primer, and hence only one strand, is labeled and the exposed film is easier to interpret than if both strands are labeled.

PCR products are size separated on 6% denaturing polyacrylamide gels and analyzed after exposure to an X-ray-sensitive film. Normally, overnight exposure of 12 to 18 hr is sufficient. A typical result is given in Figure 2.3. As a size marker, an M13 phage was sequenced with an end-labeled M13 forward primer (not shown). This allows the determination of the sizes of the alleles by comparison with the M13 sequence. Their size, minus the size of the PCR product if no repeats were present, is the number of additional bases. This number divided by three is the number of repeats in the allele (see Table 2.2). No sequence variation has been described in the sequences adjacent to the CAG repeats except in HD. In HD, a CCG repeat immediately 3′ to the CAG repeat is polymorphic with 8 alleles, although >95% of the alleles have 7 to 10 repeats. The HD chromosomes are in strong linkage disequilibrium with the most frequent $(CCG)_7$ allele, and this allele is found in >99% of the HD chromosomes.[86,87] It is therefore advisable to check the CCG repeat length (as described in the legend of Table 2.2), in addition to the CAG repeat length in patients with borderline CAG repeat sizes of their putative HD chromosomes.

2. Prenatal diagnosis

In principle, the same techniques can be used for prenatal screening. Somatic stability in chorion villi samples as compared to other fetal tissues has been demonstrated in two fetuses with an expanded HD CAG repeat,[88] but no data are available from other CAG repeat disorders.

C. Detection of the CTG Repeat in Myotonic Dystrophy

1. Polymerase chain reaction

Polymerase chain reaction allows for the detection of normal-sized and small amplified alleles. Amplification of the CTG repeats can be achieved in 20-µl volumes of reaction mixture containing 3 pmol of each primer (Table 2.2); 20 mM Tris-HCl, pH 8.4; 50 mM KCl; 1.5 mM MgCl$_2$; 0.05% W-1 detergent (GIBCO); 0.2 mM each dATP, dGTP, dCTP, and dTTP; 1 unit Taq polymerase; 100 to 300 ng template DNA. Overlay this mixture with mineral oil. Thermal cycling may be carried out for 35 cycles after an initial denaturation step of 95°C for 7.5 min, consisting of denaturation at 95°C for 1 min, annealing for 1 min at 70°C for the first 5 cycles, at 69°C for the second 5 cycles, and at 68°C for the remaining 25 cycles,

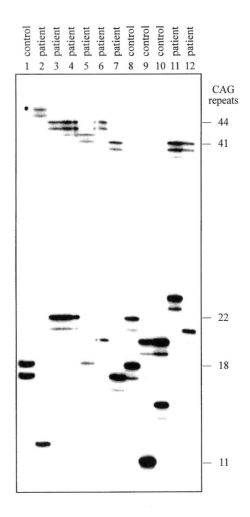

FIGURE 2.3

Molecular diagnosis of HD: Lanes 1 to 12 represent PCR products of primers Hu3 and Hu4 on a denaturing polyacrylamide gel of 8 HD patients and 4 controls. Allele sizes are 17/18 (lane 1), 12/46 (lane 2), 22/44 (lane 3), 22/44 (lane 4), 18/42 (lane 5), 20/44 (lane 6), 17/41 (lane 7), 18/22 (lane 8), 11/20 (lane 9), 15/20 (lane 10), 24/41 (lane 11), and 21/41 (lane 12).

with an extension step at 72°C for 1.5 min, ended by a final extension step at 72°C for 10 min. Primers can be end-labeled, or radioactive label can be incorporated in the PCR product as described for the PCR detection of CAG repeats. PCR products are run on 6% denaturing polyacrylamide gels and visualized by exposure to an X-ray-sensitive film. Normal-sized alleles as well as small amplified alleles up to 100 to 150 repeats become visible. The presence of two alleles in the normal size range demonstrates the absence of MD; detection of a larger allele confirms MD diagnosis.

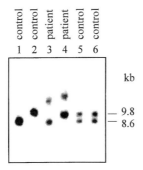

FIGURE 2.4
Detection of the CTG repeat in the DM gene after hybridization of an *Eco*RI blot with cDNA25. The controls demonstrate the deletion polymorphism with two alleles of 9.8 and 8.6 kb, respectively. Lanes 3 and 4 are DM patients with a large expanded repeat.

Large amplifications cannot be detected by PCR, and only a single normal-sized allele will become visible, as is the case for individuals homozygous for the normal-sized alleles. To detect alleles with larger expansions, Southern blot is essential.

2.   Southern blot detection

Southern blot is used to detect the expanded MD alleles of more than 100 repeats. Large repeats (>1 kb) can be visualized as bands or smears on an *Eco*RI blot, after hybridization with probe cDNA25.[89] The two 8.6- and 9.8-kb bands that may be observed in control individuals represent two alleles of a frequent *Eco*RI polymorphism at the DM locus, with allele frequencies of 0.47 and 0.53, respectively.[89,90] The disease allele is in complete linkage disequilibrium with the 9.8-kb fragment, and thus the 9.8 kb allele is always expanded in affected patients. This is illustrated in Figure 2.4. To detect extensive smears, it is advisable to make an additional long exposure of each film. Smaller amplification can better be detected after *Bam*HI digestion, and as most patients are highly mosaic in their lymphocytes, a diffuse smear that may be difficult to detect will result after hybridization of a *Bam*HI blot. Size separation and transfer to a membrane are performed as described for the Southern blot detection of fragile X syndrome. Hybridization is performed for 12 to 18 hr with radioactively labeled probe cDNA25 at 65°C. The membrane is washed for 15 min twice in 3× SSC, 0.1% SDS; twice in 1× SSC, 0.1% SDS; twice in 0.3× SSC, 0.1% SDS. SDS is removed by a brief rinse in 2× SSC. The blot is sealed in plastic foil and exposed to an X-ray-sensitive film with an intensifying screen at −80°C for 24 to 72 hr.

3.   Prenatal diagnosis

In principle, the same techniques can be used for prenatal screening.

D.   Estimating Repeat Length

Estimating the repeat length is normally performed by comparison with an M13 sequencing ladder or equivalent to estimate the repeat size. It should be noted, however, that the PCR products and the M13 sequencing ladder have a different mobility in the gel because of their reiterated nature and unequal G + C vs. A + T ratio of the former. The size of the PCR product, therefore, may not exactly correspond to the size of the corresponding M13 fragment. This difference in mobility is dependent on the gel conditions and may explain in part why differences in repeat size are sometimes observed when two laboratories analyze the same samples. Inclusion of one or two samples with the allele sizes determined by direct sequencing on each gel for reference purposes is therefore recommended. As in Table 2.2, most authors define the number of direct CGG, CAG, or CTG triplets (excluding flanking repeats as CAA, but including occasional interspersions of this sequence) as the repeat length. A few authors, however, also included adjacent triplets to a CAG repeat that encode for glutamine in their determination of the repeat length. Obviously, accurate size determination is particularly important when the allele size is in the gray zone between normal and disease alleles.

IV.   Summary

Eleven disorders and/or fragile sites caused by the expansion of nine trinucleotide repeats have been described. In the normal population, these repeats are small, polymorphic, and stably inherited. In patients, an increase of the repeat size above a threshold level causes the disorder and/or the fragile site. Greater repeat lengths cause a more severe disease and/or onset at a younger age. The enlarged repeat of patients is unstably transmitted, and the repeat length increases (or occasionally decreases) in size when transmitted to the next generation; therefore, older generations of a family affected by any dynamic disorder are generally less severely affected or are affected at a later age of onset than the younger generations, a phenomenon called anticipation. New mutations are rare, giving rise to very extended pedigrees in which the dynamic mutation segregates. As mutations other than triplet expansion have only been described for a few fragile X patients, determination of the number of repeats from the suspected disease gene is conclusive. Methods to detect these dynamic mutations were described here. The massive triplet expansions found in fragile X syndrome and myotonic dystrophy are easily detected using Southern blotting, whereas the more limited expanded alleles in the CAG-associated neurodegenerative disorders can conveniently be detected by PCR (see Table 2.3). Other methods, such as a protein detection test, have been described for the fragile X syndrome only.

TABLE 2.3
Current Detection Methods for the Three
Different Types of Triplet Repeats

	PCR Analysis	Southern Blotting	Protein Quantification
CGG repeats	Normal alleles and small premutations	Full mutations and large premutations; methylation status	Only in fragile X syndrome
CAG repeats	Normal and amplified repeats	NA	NA
CTG repeats	Normal alleles and small amplifications	Large repeats	NA

Note: NA = not applicable

Acknowledgments

We thank Dr. Rob Willemsen for details on the FMRP detection protocol and Dr. Hans Scheffer for sharing unpublished data and protocols. Edwin Reyniers, Ingrid Handig, and Lieve Vits are acknowledged for their valuable contributions to the protocols.

References

1. Verkerk, A. J. M. H., Pieretti, M., Sutcliffe, J. S., Fu, Y.-H., Kuhl, D. P. A., Pizzuti, A., Reiner, O., Richards, S., Victoria, M. F., Zhang, F., Eussen, B. E., Van Ommen, G. J. B., Blonden, L. A. J., Riggins, G. J., Chastain, J. L., Kunst, C. B., Glijaard, H., Caskey, C.T., Nelson, D. L., Oostra, B. A., and Warren, S. T., *Cell,* 65, 905–914, 1991.

2. La Spada, A. R., Wilson, E. M., Lubahn, D. B., Harding, A. E., and Fishbeck, K. H., *Nature,* 352, 77–79, 1991.

3. Richards, R. I. and Sutherland, G. R., *Cell,* 70, 709–712, 1992.

4. Knight, S. L. J., Flannery, A. V., Hirst, M. C., Campbell, L., Christodoulou, Z., Phelps, S. R., Pointon, J., Middleton-Price, H. R., Barnicoat, A., Pembrey, M. E., Holland, J., Oostra, B. A., Bobrow, M., and Davies, K. E., *Cell,* 74, 127–134, 1993.

5. Jones, C., Penny, L., Mattina, T., Yu, S., Baker, E., Voullaire, L., Langdon, W. Y., Sutherland, G. R., Richards, R. I., and Tunnacliffe, A., *Nature,* 376, 145–149, 1995.

6. Hirst, M. C., Barnicoat, A., Flynn, G., Wang, Q., Daker, M., Buckle, V. J., Davies, K. E., and Bobrow, M., *Hum. Mol. Genet.,* 2, 197–200, 1993.

7. Parrish, J. E., Oostra, B. A., Verkerk, A. J. M. H., Richards, C. S., Reynolds, J., Spikes, A. S., Shaffer, L. G., and Nelson, D. L., *Nature Genet.,* 8, 229–235, 1994.

8. Nancarrow, J. K., Kremer, E., Holman, K., Eyre, H., Doggett, N. A., Le Paslier, D., Callen, D. F., Sutherland, G. R., and Richards, R. I., *Science*, 264, 1938–1941, 1994.

9. The Huntington's Disease Collaborative Research Group, *Cell*, 72, 971–983, 1993.

10. Orr, H. T., Chung, M.-Y., Banfi, S., Kwiatkowski, T. J., Servadio, A., Beaudet, A. L., McCall, A. E., Duvick, L. A., Ranum, L. A. P. W., and Zoghbi, H. Y., *Nature Genet.*, 4, 221–226, 1993.

11. Kawaguchi, Y., Okamoto, T., Taniwaki, M., Aizawa, M., Inoue, M., Katayama, S., Kawakami, H., Nakamura, S., Nishimura, M., Akiguchi, I., Kimura, J., Narumiya, S., and Kakizuka, A., *Nature Genet.*, 8, 221–228, 1994.

12. Schöls, L., Vieira-Saecker, A. M. M., Schöls, S., Przuntek, H., Epplen, J. T., and Reiss, O., *Hum. Mol. Genet.*, 4, 1001–1005, 1995.

13. Koide, R., Ikeuchi, T., Ondonera, O., Tanaka, H., Igarashi, S., Endo, K., Takahashi, H., Kondo, R., Ishikawa, A., Hayashi, T., Saito, M., Tomod, A., Miike, T., Naito, H., Ikuto, F., and Tsuji, S., *Nature Genet.*, 6, 9–13, 1994.

14. Nagafuchi, S., Yanagisawa, H., Sato, K., Shirayama, T., Ohsaki, E., Bundo, M., Takeda, T., Tadokoro, K., Kondo, I., Murayama, M., Tanaka, Y., Kikushima, H., Umino, K., Kurosawa, H., Furukawa, T., Nihei, K., Inuoue, T., Sano, A., Komure, O., Takahashi, M., Yoshizawa, T., Kanazawa, I., and Yamada, M., *Nature Genet.*, 6, 14–18, 1994.

15. Burke, J. R., Wingfield, M. S., Lewis, K. E., Roses, A. D., Lee, J. E., Hulette, C., Pericak-Vance, M. A., and Vance, J. M., *Nature Genet.*, 7, 521–524, 1994.

16. Brook, J. D., McCurrach, M. E., Harley, H. G., Buckler, A. J., Church, D., Aburatani, H., Hunter, K., Stanton, V. P., Thirion, J. P., Hudson, T., Sohn, R., Zemelman, B., Snell, R. G., Rundle, S. A., Crow, S., Davies, J., Shelbourne, P., Buxton, J., Jones, C., Juvonen, V., Johnson, K., Harper, P. S., Shaw, D. J., and Housman, D. E., *Cell*, 68, 799–808, 1992.

17. Goldberg, Y. P., Kremer, B., Andrew, S. E., Theilmann, J., Graham, R. K., Squitieri, F., Telenus, H., Adam, S., Sajoo, A., Starr, E., Heiiberg, A., Wolff, G., and Hayden, M. R., *Nature Genet.*, 3, 174–179, 1993.

18. Myers, R. H., MacDonald, M. E., Koroshetz, W. J., Duyao, M. P., Ambrose, C. M., Taylor, S. A. M., Barnes, G., Sriinidh, J., Lin, C. S., Whaley, W. L., Lazzarini, A. M., Schwarz, M., Wolff, G., Bird, E. D., Vonsattel, J.-P.G., and Gusella, J. F., *Nature Genet.*, 3, 168–173, 1993.

19. Fu, Y.-H., Kuhl, D. P. A., Pizutti, A., Pieretti, M., Sutcliffe, J. S., Richards, S., Verkerk, A. J. M. H., Holden, J. J. A., Fenwick Jr., R. G., Warren, S. T., Oostra, B. A., Nelson, D. L., and Caskey, C. T., *Cell*, 67, 1047–1058, 1991.

20. Duyao, M., Ambrose, C., Myers, R., Novelletto, A., Perschett, F., Frontali, M., Folstein, S., Ross, C., Franz, M., Abbott, M., Gray, J., Conneally, P, Young, A., Penney, J., Hollingsworth, Z., Shoulson, I., Lazzariini, A., Falek, A., Koroshetz, W., Sax, D., Bird, E., Vonsattel, J., Bonilla, E., Alvir, J., Bckham Conde, Cha, J.-H., Dure, L., Gomez, F., Ramos, M., Sanchez-Ramos, J., Snodgrass, S., de Young, M., Wexler, N., Moscowitz, C., Penchaszadeh, G., MacFarlane, H., Anderson, M., Jenkins, B., Srinidhi, J., Barnes, G., Gusella, J., and MacDonald, M., *Nature Genet.*, 4, 387–392, 1993.

21. La Spada, A. R., Roling, D. B., Harding, A. E., Warner, C. L., Speigel, R., Hausmanowa-Petrusewicz, I., Yee, W.-C., and Fischbeck, K. H., *Nature Genet.*, 2, 301–304, 1992.

22. Tsilfidis, C., MacKenzie, A. E., Mettler, G., Barceló, J., and Korneluk, R. G., *Nature Genet.*, 1, 192–195, 1992.

23. Sherman, S. L., Morton, N. E., Jacobs, P. A., and Turner, G., *Ann. Hum. Genet.*, 48, 21–37, 1984.

24. Sherman, S. L., Jacobs, P. A., Morton, N. E., Froster-Iskenius, U., Howard-Peebbles, P. N., Nielson, K. B., Partington, N. W., Sutherland, G. R., Turner, G., and Watson, M., *Hum. Genet.*, 69, 3289–3299, 1985.

25. Reyniers, E., Vits, L., De Boulle, K., Van Roy, B., VanVelzen, D., de Graaf, E., Verkerk, A. J. M. H., Jorens, H. Z. J., Darby, J. K., Oostra, B. A., and Wilems, P. J., *Nature Genet.*, 4, 143–146, 1993.

26. Willems, P. J., Van Roy, B., De Boulle, K., Vits, L., Reyniers, E., Beck, O., Dumon, J. E., Verkerk, A. J. M. H., and Oostra, B. A., *Hum. Mol. Genet.*, 1, 511–515, 1992.

27. Turner, G., Webb, T., Wake, S., and Robinson, M., *Am. J. Med. Genet.*, 64, 196–197, 1996.

28. Hagerman, R. J., in *Fragile X Syndrome; Diagnosis, Treatment, and Research,* Hagerman, R. J. and Silverman, A. C., Eds., Johns Hopkins University Press, Baltimore, MA, 1991, pp. 1–68.

29. Heitz, D., Rousseau, F., Devys, D., Saccone, S., Abderrahim, H., Le Paslier, D., Cohen, D., Vincent, A., Toniolo, D., Della Valle, G., Johnson, S., Schlessinger, D., Oberlé, I., and Mandel, J. L., *Science*, 251, 1236–1239, 1991.

30. Oberlé, I., Rousseau, F., Heitz, D., Kretz, C., Devys, D., Hanauer, A., Boué, J., Bertheas, M. F., and Mandel, J. L, *Science,* 252, 1097–1102, 1991.

31. Yu, S., Pritchard, M., Kremer, E., Lynch, M., Nancarrow, J., Baker, E., Holman, K., Mulley, J. C., Warren, S. T., Schlessinger, D., Sutherland, G. R., and Richards, R. I., *Science,* 252, 1179–1181, 1991.

32. Kremer, E. J., Pritchard, M., Lynch, M., Yu, S., Holman, K., Baker, E., Warren, S. T., Schlessinger, D., Sutherland, G. R., and Richards, G. R., *Science,* 252, 1711–1714, 1991.

33. Reiss, A. L., Kazazian, Jr., H. H., Krebs, C. M., McAughan, A., Boehm, C. D., Abrams, M. T., and Nelson, D. L., *Hum. Mol. Genet.*, 3, 393–398, 1994.

34. Hagerman, R. J., Hull, C. E., Safandra, J. F., Carpenter, I., Staley, L. W., O'Connor, R. A., Seydel, C., Mazzocco, M. M. M., Snow, K., Thibodeau, S. N., Kuhl. D., Nelson, D. L., Caskey, C. T., and Taylor, A. K., *Am. J. Med. Genet.*, 51, 298–308, 1994.

35. Pieretti, M., Zhang, F., Fu, Y.-H., Warren, S. T., Oostra, B. A., Caskey, C. T., and Nelson, D. L., *Cell*, 66, 817–822, 1991.

36. Verheij, C., Bakker, C. E., de Graaff, E., Keulemans, J., Willemsen, R., Verkerk, A. J. M. H., Galjaard, H., Reusser, A. J. J., Hoogeveen, A. T., and Oostra, B. A., *Nature,* 363, 722–724, 1993.

37. Feng, Y., Lakkis, D., and Warren, S. T., *Am. J. Hum. Genet.*, 56, 106–113, 1995.
38. Smeets, H. J. M., Smits, A. P. T., Verheij, C. E., Theelen, J. P. G., Willemsen, R., Hoogeveen, A. T., Oosterwijk, J. C., and Oostra, B. A., van de Burgt, I., *Hum. Mol. Genet.*, 4, 2103–2108, 1995.
39. De Boulle, K., Verkerk, A. J. M. H., Reyniers, E., Vits, L., Hendrickx, J., van Roy, B., van den Bos, F., de Graaf, E., Oostra, B. A., and Willems, P. J., *Nature Genet.*, 3, 31–35, 1993.
40. Lugenbeel, K. A., Peier, A. M., Carson, N. L., Chudley, A. E., and Nelson, D. L., *Nature Genet.*, 10, 483–485, 1995.
41. Verheij, C., de Graaff, E., Bakker, C. E., Willemsen, R., Willems, P. J., Meijer, N., Galjaard, H., Reuser, A. J. J., Oostra, B. A., and Hoogeveen, A. T., *Hum. Mol. Genet.*, 4, 895–901, 1995.
42. Willems, P. J., *Nature Genet.*, 8, 213–215, 1994.
43. Sutherland, G. R. and Baker, E., *Hum. Mol. Genet.*, 1, 111–113, 1992.
44. Knight, S. J. L., Voelckel, M. A., Hirst, M. C., Flannery, A. V., Moncla, A., and Davies, K. E., *Am. J. Hum. Genet.*, 55, 81–86, 1994.
45. Mulley, J. C., Yu, S., Loesch, D. Z., Hay, D. A., Donnelly, A., Gedeon, A. K., Carbonell, P., López, I., Glover, G., Gabarrón, I., Yu, P. W. L., Baker, E., Haan, E. A., Hockey, A., Knight, S. J. L., Davies, K. E., Richards, R. I., and Sutherland, G. R., *J. Med. Genet.*, 32, 162–169, 1995.
46. Hamel, B. C. J., Smits, A. P. T., de Graaff, E., Smeets, D. F. C. M., Schoute, F., Eussen, B. H. J., Knight, S. J. L, Davies, K. E., Assman-Hulsmans, C. F. C. H., and Oostra, B. A., *Am. J. Hum. Genet.*, 55, 923–931, 1995.
47. Gedeon, A. K., Keinänen, Adès, L. C., Kääriäinen, H., Gécz, J., Baker, E., Sutherland, G. R., and Richards, R. I., *Am. J. Hum. Genet.*, 56, 907–914, 1995.
48. Nancarrow, J. K., Holman, K., Mangelsdorf, M., Hori, T., Denton, M., Sutherland, G. R., and Richards, R.I., *Hum. Mol. Genet.*, 4, 367–372, 1995.
49. Schinzel, A., auf der Maur, P., and Moser, H., *J. Med. Genet.*, 14, 438–444, 1977.
50. Kennedy, W. R., Alter, M., and Sung, J. H., *Neurology*, 18, 671–680, 1968.
51. Arbizu, T., Santamaria, J., Gomez, J. M., Quilez, A., and Serra, J. P., *J. Neurol. Sci.*, 59, 371–382, 1983.
52. Igarashi. S., Tanno, Y., Ondonera, O., Yamazaki, S., Sato, S., Ishikawa, A., Miyatani, N., Ishikawa, Y., Sahashi, K., Ibi, T., Miyatake, T., and Tsuji, S., *Neurology*, 42, 2300–2302, 1992.
53. Doyu, M., Sobue, G., Mukai, E., Kachi, T., Yasuda, T., Mitsuma, T., and Takahashi, A., *Ann. Neurol.*, 32, 707–710, 1992.
54. La Spada, A. R., Roling, D. B., Harding, A. E., Warner, R., Spiegel, R., Hausmanowa-Petrusewicz, I., Yee, W.-C., and Fishbeck, K. H., *Nature Genet.*, 2, 301–304, 1992.
55. Biancala, V., Serville, F., Pommier, J., Julien, J., Hanauer, A., and Mandel, J. L., *Hum. Mol. Genet.*, 1, 255–258, 1992.
56. Ferlini, A., Patrosso, M. C., Guidetti, D., Merlini, L., Uncini, A., Ragno, M., Plasmati, R., Fini, S., Repetto, M., Vezzoni, P., and Forabosco, A., *Am. J. Med. Genet.*, 55, 105–11, 1995.

57. Gusella, J. F., Wexler, N. S., Conneally, P. M., Naylor, S. L., Anerson, M. A., Tanzi, R. E., Watkins, P. C., Ottina, K., Wallace, M. R., Sakaguchi, A. Y., Young, A. B., Shoulson, I., Bonilla, E., and Martin, J. B., *Nature,* 306, 234–238, 1983.

58. Harper, P. S., Ed., *Huntington's Disease,* Saunders, London, 1991.

60. Andrew, S. E., Goldberg, Y. P., Thellman, J., Zeisler, J., and Hayden, M. R., *Hum. Mol. Genet.,* 3, 65–67, 1994.

61. Snell, R. G., MacMillan, J. C., Cheadle, J. P., Fenton, I., Lazarou, L. P., Davies, P., MacDonald, M. E., Gusella, J. F., Harper, P. S., and Shaw, D. J., *Nature Genet.,* 4, 393–397, 1993.

62. Andrew, S. E., Goldberg, Y. P., Kremer, B., Telenius, H., Theilmann, J., Adam, S., Starr, E., Squitieri, F., Lin, B., Kalchman, M. A., Graham, R. K., and Hayden, M. R., *Nature Genet.,* 4, 398–403, 1993.

63. Wexler, N. S., Young, A. B., Tanzi, R. E., Travers, H., Starosta-Rubenstein, S., Penney, J. B., Snodgrass, S. R., Shoulson, I., Gomez, F., Ramos-Arroyo, M. A., Penchaszadeh, G., Moreno, R., Gibbons, K., Faryniarz, A., Hobbs, W., Anderson, M. A., Bonilla, E., Conneally, P. M., and Gusella, J. F., *Nature,* 326, 194–197, 1987.

64. Kremer, B., Goldberg, P., Andrew, S. E., Theilmann, J., Telenius, H., Zeisler, J., Squitieri, F., Lin, B., Basset, A., Almqvist, E., Bird, T. D., and Hayden, M. R., *N. Engl. J. Med.,* 20, 330, 1401–1406, 1994.

65. Banfi, S., Servadio, A., Chung, M., Kawiatkowski, Jr., T. Y., McCall, A. E., Duvick, L., Shen, Y., Roth, E. J., Orr, H. T., and Zoghbi, H. Y., *Nature Genet.,* 7, 513–520, 1994.

66. Zoghbi, H. Y., *Current Neurology,* Appel, S. H., Ed., Mosby-Year Book, St. Louis, 1991, pp. 121–144.

67. Chung, M.-Y., Ranum, L. P. W., Duvick. L. A., Servadio, A., Zoghbi, H. Y., and Orr, H. T., *Nature Genet.,* 5, 254–258, 1994.

68. Ranum, L. P. W., Chung, M.-Y., Banfi, S., Bryer, A., Schut, L. J., Ramesar, R., Duvick, L. A., McCall, A., Subramony, S. H., Goldfarb, L., Gomez, C., Sandkuyl, L. A., Orr, H. T., and Zoghbi, H. Y., *Am. J. Hum. Genet.,* 55, 244–252, 1994.

69. Naito, H. and Oyanagi, S., *Neurology,* 32, 798–807, 1982.

70. Nagafuchi, S., Yanagisawa, H., Ohsaki, E., Shirayama, T., Tadokoro, K., and Yamada, M., *Nature Genet.,* 8, 177–181, 1994.

71. Hirayama, K., Takaynagi, T., Nakamura, R., Yanagiswa, N., Hattori, T., Kita, K., Yanagimoto, S., Fujita, M., Nagaoka, M., Satomura, Y., Sobue, I., Iizuka, R., Toyoura, Y., and Satoyoshi, E., *Acta Neurol. Scand.,* 89(Suppl. 53), 1–22, 1994.

72. Nørremølle, A., Nielson, J. E., Sørensen, S. A., and Hasholt, L, *Hum. Genet.,* 95, 313–318, 1995.

73. Farmer, T. W., Wingfield, M. S., Lynch, S. A., Vogel, F. S., Hulette, C., Katchnoff, B., and Jacobson, P. L., *Arch. Neurol,* 46, 774–779, 1989.

74. Rosenberg, R. N., *Mov. Dis.,* 3, 193–203, 1992.

75. Takiyama, Y., Nishizawa, M., Tanaka, H., Kawashima, S., Sakamoto, H., Karube, Y., Shimazaki, H., Soutome, M., Endo, K., Ohta, S., Kagawa, Y., Kanazawa, I., Mizuno, Y., Yoshida, T., Yuasa, Y., Horikawa, Y., Oyanagi, K., Nagai, H., Kondo, T., Inuzuka, T., Onodera, O., and Tsuji, S., *Nature Genet.,* 4, 300–304, 1993.

76. Stevanin, G., Le Guern, E., Ravisé, N., Chneiweiss, H., Dürr, A., Cancel, G., Vignal, A., Boch, A.-L., Ruberg, M., Penet, C., Pothin, Y., Lagroua, I., Hagenau, M., Rancrel, G., Weissenback, J., Agid, Y., and Brice, A., *Am. J. Hum. Genet.,* 54, 11–20, 1994.

77. Scheffer, H., personal communication

78. Harper, P. S., Ed., *Myotonic Dystrophy*, Saunders, London, 1989.

79. Pergolizzy, R. G., Erster, S. H., Goonewardena, P., Brown, W. T., *Lancet,* 339, 271–272, 1992.

80. Oostra, B. A., Willems, P. J., and Verkerk, A. M. J. H., in *Genome Maps and Neurological Disorders,* Davies, K. E. and Tilghman, S. M., Eds., Cold Spring Harbor Laboratory Press, Cold Spring Harbor, NY, 1993, pp. 45–75.

81. Sutherland, G. R., Gegeon, A., Komman, L., Donnelly, A., Byard, R. W., Mulley, J. C., Kremer, E., Lynch, M., Pritchard, M., Yu, S., and Richards, R. I., *N. Engl. J. Med.,* 325, 1720–1722, 1991.

82. Rousseau, F., Heitz, D., Biancalana, V., Blumenfeld, S., Kretz, C., Boué, Tommerup, N., Van Der Hagen, C., DeLozier-Blanchet, Croquette, M.-F., Gilgenkrantz, S., Jalbert, P., Voelckel, M.-F., Oberlé, I., and Mandel, J.-L., *N. Engl. J. Med.,* 325, 1673–1681, 1991.

83. Willemsen, R., Mohkamsing, S., De Vries, B., Devys, D., van den Ouweland, A., Mandel, J.-L., Galjaard, H., and Oostra, B. A., *Lancet,* 345, 1147–1148, 1995.

84. Devys, D., Lutz, Y., Rouyer, N., Bellocq, J.-P., and Mandel, J.-L., *Nature Genet.,* 4, 335–340, 1993.

85. Sutcliffe, J. S., Nelson, D. L., Zhang, F., Pieretti, M., Caskey, C. T., Saxe, D., and Warren, S. T., *Hum. Mol. Genet.,* 1, 397–400, 1992.

86. Barron, L. H., Rae, A., Holloway, S., Brock, D. J. H., and Warner, J. P., *Hum. Mol. Genet.,* 3, 173–175, 1994.

87. Rubinsztein, D. C., Amos, W., Leggo, J., Goodburn, S., Ramesar, R. S., Old, J., Bontrop, R., McMahon, R., Barton, D. E., and Ferguson-Smith, M. A., *Nature Genet.,* 7, 525–530, 1994.

88. Benitez, J., Robledo, Ramos, C., Ayuso, C., Astarloa, R., Garcia, J., and Brambati, B., *Hum. Genet.,* 96, 229–232, 1995.

89. Buxton, J., Shelbourne, P., Davies, J., Jones, C., Van Tongeren, T., Aslanidis, C., de Jong, P., Jansen, G., Anvret, M., Riley, B., Williamson, R., and Johnson, K., *Nature,* 355, 547–548, 1992.

90. Harley, H. G., Brook, J. D., Rundle, S. A., Crow, S., Reardon, W., Buckler, A. J., Harper, P. S., Housman, D. E., and Shaw, D. J., *Nature,* 355, 545–546, 1992.

91. Riess, O., Noerremoelle, A., Soerensen, S. A., and Epplen, J. T., *Hum. Mol. Genet.,* 2, 637, 1993.

92. Li, S.-H., McInnis, M. G., Magolis, R. L., Antonarakis, S. E., and Ross, C. A., *Genomics,* 16, 572–579, 1993.

93. Mahadevan, M., Tsilfidis, Sabournin, L., Shutler, G., Amemiya, C., Jansen, G., Neville, C., Narang, M., Barceló, O'Hoy, K., Leblond, S., Earle-Macdonald, J., de Jong, P. J., Wieringa, B., and Korneluk, R. G., *Science,* 255, 1253–1255.

94. Rubinsztein, D. C., Leggo, J., Barton, D. E., and Ferguson-Smith, M. A., *Nat. Genet.,* 5, 214–215, 1993.

95. Hirst, M., Grewal, P., Flannery, A., Slatter, R., Maher, E., Barton, D., Fryns, J.-P., and Davies, K., *Am J. Hum. Genet.,* 56, 67–74, 1995.

96. Gu, Y., Lugenbeel, K. A., Vockley, J. G., Grody, W. W., and Nelson, D. L., *Hum. Mol. Genet.,* 3, 1705–1706, 1994.

97. Wöhrle, D., Kotzot, D., Hirst, M. C., Manca, A., Korn, B., Schmidt, A., Barbi, G., Rott, H.-D., Poustka, A., Davies, K. E., and Steinback. P., *Am. J. Hum. Genet.,* 51, 299–306, 1992.

98. Trottier, Y., Imbert, G., Poustka, A., Fryns, J.-P., and Mandel, J.-L., *Am. J. Med. Genet.,* 51, 454–457, 1994.

99. Meijer, H., de Graff, E., Merckx, D. M. L., Jongbloed, R. J. E., de Die-Smulders, C. E. M., Engelen, J. J. M., Fryns, J.-P., Curfs, P. M. G., and Oostra, B. A., *Hum. Mol. Genet.,* 3, 615–620, 1994.

100. Quan, F., Zonana, J., Gunter, K., Peterson, K. L., Magenis, R. E., and Popovich, B. W., *Am. J. Hum. Genet.,* 56, 1042–1051, 1995.

101. Albright, S. G., Lachiewicz, A. M., Tarleton, J. C., Rao, K. W., Schwartz, C. E., Richie, R., Tennison, M. B., and Aylsworth, A. S., *Am. J. Med. Genet.,* 51, 294–297, 1994.

102. Gedeon, A. K., Baker, E., Robinson, H., Partiington, M. W., Gross, B., Manca, A., Korn, B., Poustka, A., Yu, S., Sutherland, G. R., and Mulley, J. W., *Nature Genet.,* 1, 341–344, 1992.

103. Clarke, J. T. R., Wilson, P. J., Morris, C. P., Hopwood, J. J., Richards, R. I., Sutherland, G. R., and Ray, P. N., *Am. J. Hum. Genet.,* 51, 316–322, 1992.

104. Dahl, N., Hu, L.-J., Chery, M., Gilgenkrantz, S., Nivelon-Chevallier, A., Gouyon, J.-B., and Mandel, J.-L., *Cytogen. Cell Genet.,* 64, 181, 1993.

105. Schmidt, M., *Am. J. Med. Genet.,* 43, 279–281, 1992.

Chapter **3**

Gene Targeting via Triple Helix Formation

Gan Wang and Peter M. Glazer

Contents

I.   Introduction

Triplex DNA can be formed when oligonucleotides bind as third strands of DNA in a sequence-specific manner, fitting into the major groove of the double helix in homopurine/homopyrimidine stretches in duplex DNA.[1-3] Progress in elucidating the third-strand binding code has raised the possibility of developing nucleic acids as sequence-specific DNA binding reagents for research and possibly clinical applications.

Oligonucleotide-mediated triple helix formation has been shown to prevent transcription factor binding to promoter sites and to block mRNA synthesis in a variety of target genes, including the c-*myc* in HeLa cells.[4,5] Triplex-forming oligonucleotides (TFOs) have also been designed that bind selectively to HIV-1 provirus DNA,[6] and TFOs targeted to the replication origin of SV40 virus were reported to inhibit viral replication in culture.[7]

TFOs have also been developed as tools to generate site-specific cleavage of target DNA. Strobel et al.[8] used a 16-mer TFO to direct enzymatic cleavage of human chromosome 4 at a single site. TFOs directly linked to reactive reagents have also been used in experiments to generate site-specific chemical cleavage of DNA. For example, Moser et al.[2] used a 19-mer TFO conjugated to EDTA-Fe, which cleaves the DNA via radical-mediated oxidation. Perrouault et al.[9] employed an 11-mer covalently attached to an ellipticine derivative to carry out site-specific photo-induced DNA cleavage.

As another potential application, we reasoned that it would be advantageous to use triple helix formation to target mutations to specific sites in selected genes in order to produce permanent, heritable changes in gene function and expression.[10-12] In this approach, mutations are targeted to a selected site by linking the triplex-forming oligonucleotide to a mutagen so that the sequence specificity of the triplex formation can be imparted to the action of the mutagen (Figure 3.1).

As a potential research tool for inducing mutations in a selected gene, triple helix-targeted mutagenesis has several unique attributes. Although this technique

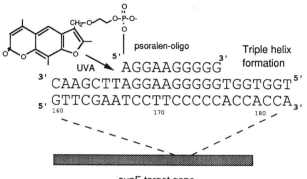

psoralen-oligo

UVA

Triple helix formation

5' AGGAAGGGGG 3'
3' CAAGCTTAGGAAGGGGGTGGTGGT 5'
5' GTTCGAATCCTTCCCCCACCACCA 3'
160   170   180

supF target gene

FIGURE 3.1

Strategy for triple helix-targeted mutagenesis. The 10-base triple helix-forming oligonucleotide is shown positioned above its target site in the *supF* gene (base pairs 167 to 176). The oligonucleotide is conjugated to a mutagen, 4'-hydroxymethyl-4,5',8-trimethylpsoralen, which is targeted to intercalate between base pairs 166 and 167, as indicated by the arrow. Photoactivation of the psoralen generates adducts and thereby mutations at the targeted site.

may be useful for generating targeted mutations in duplex DNA *in vitro,* it may have greater value as an approach to generate gene-specific mutations *in vivo*. In our recent work, we have found that triplex-forming oligonucleotides can enter mammalian cells and introduce targeted lesions and thereby mutations in a chosen gene.[12]

The same strategy could, in theory, be used to target mutations to selected genes within viruses, bacteria, or yeast. Efforts are also underway in our laboratory to use this technique on both *Caenorhabditis elegans* and *Drosophila*. Ultimately, triplex-mediated targeting may offer an alternative to standard gene transfer techniques for the genetic manipulation of cells, especially for gene knockout experiments. This would bypass the need for several complicated experimental steps, including *in vitro* site-directed mutagenesis of the chosen gene, construction of a selection vector for homologous recombination and gene replacement, transfer of that construct into the recipient cells, and selection for successful gene replacement or disruption.

Triplex-mediated gene targeting depends on the ability to design a third strand oligonucleotide to bind to a selected site in double-stranded DNA. Our current understanding of DNA triple helices and the existing collection of nucleotide analogs limits triple helix formation to regions of DNA with polypurine:polypyrimidine sequences. The general applicability of this technique will depend on the extension of the third strand binding code and on the development of nucleotide analogs so that triplex formation can occur at any sequence with the required specificity and binding affinity. Much work in this regard is underway.[13-16]

Operationally, triplex-directed mutagenesis involves using the third strand to position a mutagen at the desired location in the DNA. We have used psoralen, but other mutagens, such as alkylating agents, could be employed to produce a different spectrum of mutations. Following triplex formation and site-specific psoralen intercalation, the psoralen is photoactivated with long wavelength ultraviolet (UVA) light. This generates a targeted psoralen adduct in the DNA. A mutation is produced when this targeted premutagenic lesion is processed by cellular repair and replication activities into a mutation. Hence, the DNA must be on a replicon that can replicate in either bacteria or, preferably, mammalian cells. For *in vivo* mutagenesis, the target gene will be either on a chromosome or in a viral genome. Our work has shown that mostly T:A to A:T transversions are produced at the targeted psoralen intercalation site; however, other base pair changes and small deletions encompassing the target site are occasionally observed.

The yield and specificity of the mutagenesis depend both on the deficiency of triplex formation and on the cellular repair of the targeted damage. For example, we observed only 0.2% mutations upon replication of a damaged lambda vector in *Escherichia coli,*[11] whereas we found 7.4% mutations when a SV40 replicon containing the targeted adduct was replicated in monkey COS cells.[10]

Recently, we found that TFOs with sufficient binding affinity but without a conjugated mutagen could still induce targeted mutagenesis of a reporter gene in SV40 DNA in mammalian cells.[17] This process appeared to be a result of the ability of triple helices to block transcription and trigger a somewhat error-prone gratuitous transcription-coupled repair on an otherwise undamaged template. In other work, we found that the repair of a targeted psoralen adduct could be influenced by the presence of an associated third strand. Taken together, these results suggest that cellular DNA metabolism can be altered by triple helices. This effect could possibly be exploited to generate an unrepairable DNA lesion, perhaps leading to selective cell death or viral inactivation.

Clearly, many potential applications of triplex DNA remain speculative. Here, we discuss methods that have allowed us to achieve targeted mutagenesis of a mutation reporter gene in mammalian cells.[12] Although this work was carried out in a specialized model system, there is potential for eventual use of this technology in targeting of naturally occurring, endogenous genes in mammalian cells and viruses.

II.   Materials

A.   Oligonucleotides

Psoralen-linked oligonucleotides can be obtained from Oligos Etc. (Wilsonville, OR) or can be synthesized by standard automated methods using materials from Glen Research (Sterling, VA). The psoralen is incorporated into the oligonucleotide synthesis as a psoralen phoshoramidite, resulting in an oligonucleotide linked at the 5' end via a two-carbon linker arm to 4'-hydroxymethyl-4,5', 8-trimethylpsoralen, as illustrated in Figure 3.1. Oligonucleotides are stored at –20°C and protected from light.

B.   Vectors

SV40-based shuttle vectors, pSP189 and pSupFG1,[12] each containing the SV40 origin of replication and large T-antigen gene along with the pBR327 replication origin and β-lactamase gene, were used in our work (Figure 3.2). pSP189 contains the *supF* amber suppressor tRNA gene as a mutation reporter gene, whereas pSupFG1 contains the *supFG1* mutation reporter gene, a *supF* derivative with a modified sequence designed for strong third-strand binding. These vectors were used as models for gene-targeting experiments. The *supF* and *supFG1* genes are cloned into unique *Xho*I and *Eag*I sites in the vectors and can be replaced with another gene of interest.

C.   Cells

Monkey COS-7 cells (from the American Type Culture Collection #1651-CRL) have been used in our studies. *E. coli* SY204 (*lacZ* [amber]) was used as a host for screening *sup*F gene mutations.[12]

D.   Buffers

Triplex-binding buffer: 10 mM Tris HCl, pH 7.4; 1 mM spermidine; and 20 mM MgCl$_2$. Cell resuspension solution: 50 mM Tris-HCl; 10 mM EDTA, pH 8.0; 100 µg/ml RNase A. Cell lysis solution: 0.2 M NaOH, 1% SDS. Neutralization solution: 3 M potassium acetate, pH 5.5. TE buffer: 10 mM Tris-HCl, pH 8.0; 1 mM EDTA.

E.   Light Source

Psoralen photoactivation was accomplished using broad-band UVA light centered at 365 nm. UVA lamps are available from Southern New England Ultraviolet Co. (Bradford, CT). Dosimetry should be performed using a radiometer, such as the IL 1400 from International Light (Newburyport, MA). A typical UVA irradiance is 5 mW/cm^2.

F.   Cell-Growth Media

COS cell-growth medium, Dulbecco's modified Eagle medium supplemented with 10% fetal calf serum (FCS); available from GIBCO-BRL, Gaithersburg, MD, or another standard supplier.

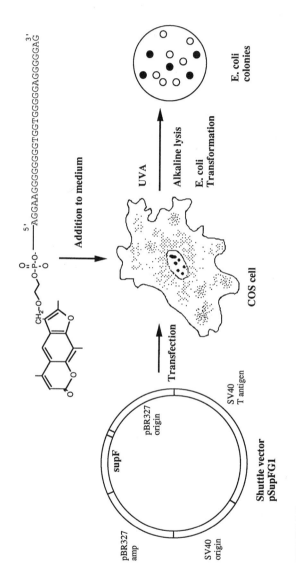

FIGURE 3.2

Experimental strategy for targeted mutagenesis of SV40 DNA mediated by intracellular triple helix formation within COS cells. The SV40 shuttle vector DNA, pSupFG1, is transfected into COS cells by electroporation. Separately, the psoralen-conjugated oligonucleotides (pso–AG30 is shown) are subsequently added to the extracellular growth medium. After allowing time for oligonucleotide entry into cells and for possible intracellular triple helix formation, the cells are exposed to UVA irradiation to photoactivate the psoralen and generate targeted adducts. Following vector replication, the cells are lysed, and the vector molecules are harvested for transformation into E. coli to allow genetic analysis of the supF gene. The structure of the tethered psoralen (4′-hydroxymethyl-4,5′,8-trimethylpsoralen attached at the 4′ hydroxymethyl position via a two-carbon linker arm to the 5′ phosphate of the oligonucleotide) is shown. (From Wang, G. et al., Mol. Cell. Biol., 15, 1759, 1995. With permission.)

G.   Bacterial Plates

Use standard LB agar supplemented with ampicillin. Add 10 g bactotryptone, 5 g bacto-yeast extract, and 10 g NaCl per liter of dH_2O. Adjust pH to 7.5 with NaOH. Add 15 g bactoagar per liter. Autoclave. Allow to cool to 55°C before adding ampicillin to a final concentration of 50 µg/ml. Ampicillin stock: 10 mg/ml of the sodium salt of ampicillin in water. Sterilize by filtration and store at –20°C.

III.   Methods

A.   Target Site Choice

Most triple helices studied so far have been formed within homopurine/homo-pyrimidine stretches. At sites of mixed sequences, few stable triple helices have been formed.[18] Pyrimidine interruptions in the homopurine strand of the duplex target destabilize potential triplex formation. Many laboratories are developing novel nucleo-side analogs for incorporation into TFOs to overcome this destabilizing effect and to allow third-strand to non-homopurine sites. Generally, however, given the current technology, the target site must be chosen to consist of polypurine:polypyrimidine sequences. Also, if psoralen-mediated mutagenesis is anticipated, the site should be chosen to have a 5′ TpA 3′ site at one end of the homopurine run as the target for psoralen photomodification. In this way, the psoralen at the end of the third strand is positioned adjacent to the TpA site, which enhances psoralen intercalation and photoreaction. If another mutagen is used, this is not required. For *in vitro* triplex formation, a homopurine/homopyrimidine stretch of 10 bp can serve as a suitable target;[10] however, to achieve any intracellular binding using standard TFOs, longer homopurine/homopyrimidine stretches (13 or more base pairs) are required.[12]

B.   Oligonucleotide Design and Synthesis

The design of a candidate TFO depends on the composition of the intended target sequence. For A:T base pair-rich target sites, a homopyrimidine oligonucleotide is preferred, with the TFO binding in a direction parallel to the purine strand in the duplex (Figure 3.3). Studies show that in the triplex, T binds to A:T pairs and C binds to G:C pairs.[1-3] For G:C base pair-rich target sites, a homopurine strand that binds anti-parallel to the purine strand in the Watson-Crick duplex is preferred, with A binding to A:T and G binding to G:C base pairs.[19] The specificity of triplex formation arises from these base triplets (A.A:T and G.G:C in the purine motif) formed by hydrogen bonding; mismatches destabilize the triple helix. Beyond these basic rules, however, some sequence variation can occur and still allow triplex formation. For example, T can substitute for A in the purine motif and bind to A:T base pairs.

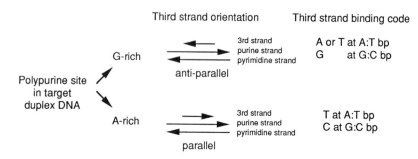

FIGURE 3.3

Guidelines for the design of triple helix-forming oligonucleotides. Triple helices form most readily at polypurine:polypyrimidine sites in DNA. If these sites are G:C base pair-rich, triplex formation in the anti-parallel motif is favored, whereas A:T base pair-rich sites are conducive to parallel triple helix formation, as indicated. The arrows indicate the 5' to 3' direction of the strands. (From Wang, G. et al., *Mol. Cell. Biol.*, 15, 1759, 1995. With permission.)

Triplex formation by T- and C-containing TFOs in the parallel pyrimidine motif is pH dependent, since it depends on protonation of C. Replacing the cytosine in the triplex strand with various analogs has been proposed as a means to eliminate the pH dependence.[14,20]

In contrast, anti-parallel, purine-motif triple helices are pH independent, forming readily at physiological pH. Consequently, it has been the purine motif that has been used most successfully in inhibiting gene expression *in vivo*[5] and in inducing targeted mutagenesis of reporter genes both *in vitro* and *in vivo*.[12]

The binding affinity of triplex-forming oligonucleotides to their target sites also depends on the length of the oligonucleotides. Generally, oligonucleotides of length 13 to 30 can form very stable triplexes (with equilibrium dissociation constants in the 10^{-8}- to 10^{-10}-M range).

C.   Triplex Binding Assays

Third-strand binding to the target duplexes can be measured by a gel mobility shift assay. For example, in the case of pso-AG30 binding to *supFG1*, two complementary 57-mers containing the sequence corresponding to base pairs 157 to 213 of *supFG1* were synthesized. One oligomer was end-labeled by using T4 polynucleotide kinase and [γ-^{32}P]ATP. Duplex DNA was prepared by mixing the [γ-^{32}P]ATP-labeled oligo with the unlabeled complementary oligo at a ratio of 1:1 in TE buffer and was incubated in the solution at 37°C for 2 hours. A fixed concentration of duplex DNA (10^{-10} M) was incubated with increasing concentrations of the psoralen-linked oligomer in 10 ml of a solution containing 20 mM MgCl$_2$, 10 mM Tris-HCl (pH 2.4), and 1 mM spermidine at 37°C for 2 hr.

UVA irradiation (1.8 J/cm² of broad-based UV light centered at 365 nm, irradiance of 5 mW/cm²) was used to generate photoadducts and thereby to link covalently the oligomers to their targets. This dose generates approximately 60 to 70% psoralen interstrand cross-links (XL, between the two strands of the target duplex, in addition to the tether connecting the psoralen to the TFO) and 25 to 35% psoralen monoadducts (MA), in the context of the triple helix.[21] The samples were mixed with 90 µl of formamide, and a 20-µl aliquot of each sample was analyzed on an 8% polyacrylamide denaturing gel containing 7 M urea. A phosphorimager (Molecular Dynamics; Sunnyvale, CA) was used for quantitation of the reaction products. The concentration at which triplex formation (as indicated by the generation of specific photoadducts) was half maximal was taken as the equilibrium dissociation constant (Kd). In the case of pso-AG10 binding to *supF*, a 24-bp synthetic duplex target, corresponding to base pairs 160 to 183 in *supF*, was constructed and end-labeled, with binding measured as described above.

D. Mutagenesis Protocols

1. Targeted mutagenesis of SV40 DNA
by triplex formation *in vitro*

The pSP189 vector DNA at 80 n*M* was incubated with pso-AG10 at 1 µ*M* in triplex binding buffer in a total volume of 10 ml. UVA (365 nm) irradiation of the samples was performed at a dose of 1.8 J/cm² The oligonucleotide-plasmid complex was then transfected into monkey COS-7 cells, using cationic liposomes (DOTAP; Boehringer Mannheim) at a final concentration of 5 µg/ml in a culture dish containing 10⁴ cells per cm². The next day, the medium containing the liposome mixture was replaced by fresh growth medium. The cells were grown for 48 hr, allowing the vector DNA to be repaired and replicated. SV40 vector DNA was harvested from the COS cells by the protocol described below.

2. Targeted mutagenesis by
triplex formation *in vivo*

The COS cells, at 70% confluence, were washed with phosphate-buffered saline (PBS)-EDTA, treated with trypsin, and incubated at 37°C for 5 min. The cells were resuspended in Dulbecco's modified Eagle medium supplemented with 10% fetal calf serum, and were washed three times by centrifugation at 900 rpm for 5 min (4°C) in a Sorvall RT6000D. The cells were finally resuspended at 10⁷ cells per ml. The plasmid DNAs were added at 3 µg of DNA per 10⁶ cells, and the cell-DNA mixture was left on ice for 10 min. Transfection of the cells was performed by electroporation with a Bio-Rad gene pulser at a setting of 25 µF, 250 Ω, and 250 V in a 0.4-cm-diameter cuvette. Following electroporation, the cells were kept on ice for 10 min. The cells were then diluted with growth medium, washed, and transferred to 37°C for 30 min. At this point, the cells were further diluted and exposed to the oligonucleotides in growth medium at 2 μM while in suspension at 37°C with gentle agitation every 15 min. UVA irradiation was administered 2 hr later at a dose of 1.8 J/cm². All samples, including control cells not exposed to oligonucleotides, received UVA irradiation. The cells were further diluted in growth medium and allowed to attach to tissue culture dishes at a density of 2 × 10⁴ cells per cm². The cells were harvested 48 hr later for vector analysis.

E.   Rescue of Viral Vectors for Analysis

The viral vector DNA was isolated from the mammalian cells by a modified alkaline lysis procedure. The cells were detached by trypsinization, washed, and resuspended in 100 μl of cell resuspension solution. An equal volume of cell lysis solution was added, followed by 100 μl of neutralization solution. A 15-min room temperature incubation was followed by centrifugation in a microcentrifuge for 10 min. The supernatant was extracted with an equal volume of pheno/chloroform (1:1) once, and the DNA was precipitated with 2.5 volumes of ethanol at –70°C for 10 min. The DNA was collected by centrifugation for 10 min, washed with 70% ethanol once, and allowed to air dry for 5 min at room temperature (RT). The DNA was digested with *Dpn*I (to eliminate unreplicated vectors which had not acquired the mammalian methylation pattern) and RNase A at 37°C for 2 hr, extracted with phenol-chloroform, and precipitated with ethanol. The DNA pellet was dissolved in 10 μl of TE buffer, and 1 μl of the sample of vector DNA was used to transform *E. coli* SY204 (*lacZ125* [amber]) by electroporation (BioRad; settings: 25 μF, 250 Ω, and 1800 V; 0.1-cm diameter cuvette).

F.   Identification and Analysis
of Mutations

The transformed *E. coli* cells were plated onto Luria-Bertani plates containing 50 μg of ampicillin per ml, 100 μg of X-Gal per ml, and 1 μM IPTG (isopropyl-β-D-thiogalactopyranoside) and were incubated at 37°C overnight. Mutant colonies containing inactivated *supF* genes unable to suppress the amber mutation in the host cell β-galactosidase gene were detected as white colonies among the wild-type blue ones. The mutant colonies and the total colonies were counted. The mutant colonies were purified, and the plasmids were isolated for DNA sequence analysis.

Alternately, plasmids that carry mutations in the target gene can be identified by several methods. Colony *in situ* hybridization with oligonucleotide probes matching the desired mutant sequence can be used to screen the mutations among a large number of wild-type colonies.[10] Allele-specific PCR also provides a very sensitive tool to detect the mutations in pooled samples. Sequence analysis can be used to localize mutation sites within the target gene.

IV. Results

A. Targeted Mutagenesis of λ DNA by *In Vitro* Triple Helix Formation

Our initial experiments investigating triplex-mediated targeted mutagenesis were performed using an intact, double-stranded λ DNA as a target. The *supF* gene was cloned into a λ vector in previous work.[22] A 10-base oligonucleotide, 5′ AGGAAGGGGG 3′ (AG10) was synthesized to bind to base pairs 167 to 176 of the *supF* gene in the anti-parallel motif (Figure 3.1). A psoralen was connected to the oligonucleotide at the 5′ adenine via a two-carbon linker arm to generate pso-AG10, as shown (Figure 3.1), with the goal of directing psoralen intercalation and mutations to base pairs 166 to 167. UVA (centered at 365 nm) irradiation was used to activate the psoralen in order to form a premutagenic adduct on the thymidines within base pairs 166 or 167.

The λ phage DNA was then used as a substrate in λ *in vitro* packaging reactions, which package the phage DNA into infectious phage particles, allowing growth of each phage as an individual plaque on a bacterial lawn. Analysis of the *supF* gene was facilitated by the use of bacteria carrying an amber nonsense mutation in the *lacZ* gene encoding the β-galactosidase enzyme. Suppression of the amber mutation in *lacZ* (amber) by a functional *supF* generates active enzyme which can metabolize the chromogenic substrate, X-gal, and turn it blue. Hence, phage with functional *supF* genes yields blue plaques on a lawn of *lacZ* (amber) bacteria in the presence of X-gal, while phage-bearing mutations in the *supF* gene fail to suppress the *lacZ* (amber) mutation and yield colorless plaques.

Pso-AG10 plus UVA treatment of the λ DNA resulted in a mutation frequency of 0.23% in *supF*, but approximately 100-fold less (0.0024%) in the lambda repressor gene (*cI*) which is also present in the phage DNA. The background in untreated λ DNA was 0.0009%. When the λ*supF* was treated with pso-GA 10 (consisting of the reverse of the AG10 sequence), the mutation frequency was 0.0004%, which is similar to the untreated DNA. Another control, a psoralen-linked oligonucleotide containing all four nucleotides (pso-TCAG10), could not induce targeted mutagenesis of *supF* genes above background. No significant mutagenesis was produced by UVA alone (1.8 J/cm^2) in the absence of the pso-AG10. This data provided genetic evidence for the targeted mutagenesis of the *supF* gene by pso-AG10.

To obtain direct evidence for targeted mutagenesis, a series of independent mutants produced in the *supF* gene of the λ vector by pso-AG10 and UVA were sequenced. All except one of the 25 mutations produced by pso-AG10 were at or near the targeted T:A base pairs at positions 166 to 167. Further 56% of the mutations consist of the same T:A to A:T transversions precisely at the targeted base pair (167), demonstrating the specificity and reproducibility of the targeting by the pso-AG10. The predominance of the T:A to A:T transversions at this site were consistent with the mutagenic action of psoralen, which tends to form adducts at pyrimidines, especially at thymidines.

B.   Targeted Mutagenesis of SV40 DNA by *In Vitro* Triplex Formation

Triplex-mediated targeted mutagenesis was studied in pSP189, an SV40-based shuttle vector which contains both the SV40 and the pBR327 origins of replication and the β-lactamase gene for ampicillin resistance.[10] The plasmid also carries the *supF* reporter gene. The pSP189 plasmid DNA was incubated with pso-AG10 *in vitro* and was exposed to UVA light at 1.8 J/cm². The DNA-pso-AG10 complex was transfected into COS cells using cationic lipids. After two days of cell growth to allow for repair and replication, the vector DNA was rescued from the cells and was used to transform *E. coli* SY204 (*lacZ* [amber]) to facilitate genetic analysis of the *supF* gene. Targeted mutations in the *supF* gene were induced by pso-AG10 at frequencies as high as 6.4%, while pso-GA10 produced essentially no mutations above background (0.07% vs. 0.06%, respectively). Out of 20 mutations generated by pso-AG10 that were analyzed, 11 were found to carry the same T:A to A:T transversions at base pair 167. These results demonstrated the possibility of achieving targeted mutagenesis of SV40 viral DNA using psoralen-linked TFOs.

C.   Targeted Adduct Formation within Mouse Genomic DNA *In Vitro*

The ability of pso-AG10 to efficiently find and bind to its target site in the *supF* gene even within the context of the complex genome of a mammalian cell, consisting of approximately 3×10^9 bp, was investigated. DNA from transgenic mice, carrying multiple (about 100) tandem copies of λ*supF* in their chromosomes, was used to assay targeted mutagenesis of mouse DNA *in vitro* in experiments parallel to those performed with pure λ DNA. The mouse DNA was incubated with either pso-AG10 or pso-GA10, and the samples were irradiated with UVA light. The DNA was used as a substrate in λ *in vitro* packaging reactions, which can identify, cut out, and package the lambda vector DNA from within the mouse DNA into viable phage particles. The resulting phage were grown in bacteria, and *supF* mutagenesis was assayed.

In these experiments, the ultimate production of mutations depends on bacterial processing of the lesions present in the packaged λ DNA rescued from the mouse DNA. The results showed that the frequency of mutations targeted to *supF* is similar to that obtained with pure λ DNA (approximately 2 per 1000). This suggests that the triplex-forming oligonucleotide can find its binding site within the mouse DNA quite efficiently, but that the yield of mutations is limited by bacterial processing. Treatment of the mouse DNA with the reverse oligomer, pso-GA10, did not yield mutagenesis above background in the *supF* gene. This is consistent with the inability of that oligomer to modify the target site in the *supF* gene. In other experiments, it was found that incubation times as short as

10 min yielded mutation frequencies similar to those obtained with 24 hr of incubation, implying that site-specific triplex formation in the mouse genome is quite rapid, at least *in vitro*. Some caution should be exercised in interpreting these results since there are approximately 100 copies of the λ genome present within the strain of transgenic mice used, and so this is not representative of a single copy gene.

D. Targeted Mutagenesis of SV40 DNA Mediated by Intracellular Triplex Formation

In all the above experiments, the triple helix formation was carried out *in vitro* on naked DNA. The utility of this approach for gene targeting, however, depends on the ability of TFOs to bind to and modify target cells within cells. We therefore investigated the ability of TFOs to mediate targeted mutagenesis via *intracellular* triplex formation, again taking advantage of the SV40 assay system (Figure 3.3).[12] For these experiments, a new vector, pSupFG1, was constructed. This vector is derivative of pSP189 containing a modified *supF* gene (*supFG1*) with a longer third-strand binding sequence, 5′ TCCTTCCCCCCCCCACCCCCTCCCCCTC 3′, at base pairs 167 to 196 at the 3′ end of the gene. A 30-base psoralen-conjugated TFO, pso-AG30, 5′ pso-AGGAAGGGGGGGGTGGTGGGGGAGGGGGAG 3′, was designed to bind to this site in an anti-parallel orientation. *In vitro* triplex binding analysis indicated that the Kd of pso-AG30 binding to *supFG1* is $3 \times 10^{-9} M$, more than 300 times stronger than that of pso-AG10 binding to either *supF* or *supFG1*.

When the COS cells pretransfected with the pSupFG1 plasmid were subsequently treated with pso-AG30 and exposed to UVA light, targeted mutagenesis of SV40 viral DNA was achieved at frequencies as high as 2.1% (with pso-AG30 at a 2-μM concentration in the extracellular medium; Figure 3.3).[12] But when the cells were treated with pso-AG10 at the same concentration, no mutations were produced above the background frequency. These results showed that targeted mutagenesis could be achieved within cells, and they indicated that oligonucleotides with strong third-strand binding affinity (Kd's in the range of $10^{-9}M$) are required to mediate significant levels of mutagenesis *in vitro*.

We also varied the time between TFO treatment of the cells and the UVA irradiation. The highest mutation frequency was found when the cells were irradiated 2 hr after TFO treatment which suggested that at least 2 hr are required for TFO uptake by the cells and for intracellular triplex formation. Inferior results were found at later times, probably due to gradual TFO degradation within the cells. Using TFOs containing mixed sequences, we also found that just a few mismatches in the triplex code eliminated the ability of the TFOs to mediate gene targeting. Of the mutations in the *supFG1* gene induced by pso-AG30, 20 were analyzed by DNA sequencing (Figure 3.4). Of 21 mutations, 15 (71%) were T:A to A:T transversions at position 166 (the predicted psoralen intercalation site). These results again demonstrated the base-pair specificity of targeting.

```
                                                                                                         A        5'
                                                                                                         A
                                                                                                         A
                                                                                                         A
                                                                                                         A
                                                                                                         A
                                                                                                         A
                                                                                                         A
                                                                                                         A
                                                                                                         A
                                                                                                         A
                                                                                                         A
                                                                                                     CG  A
3'
CATTTTCGTAATGACACCACCCCAAGGGCTCGCCGGTTTCCCTCGTCTGAGATTTAGACGGCAGTAGCTGAAGCTTCAAGCTTAGGAAGGGGGGGGTTGGTGGGGAGGGGGAG
   90       100       110       120       130       140       150       160       170       180
GTAAAAGCATTACCTGTGGTGGGGTTCCCGAGCGGCCAAAGGAGCCAGACTCTAAATCTGCCGTCATCGACTTCGAAGTTCGAATCCTTCCCCCCCACCACCCCCTCCCCCTC
5' +    ++  +++++++++++++++++++++++++++++   ++++++  ++++++  +++++++++++++++++++++++  3'
pre-tRNA (58-98) Suppressor tRNA (99-183)
                                                                          |-----(Δ bp 161-188)--------|
                                                                          |-----(Δ bp 164-217)------------>>
                                                      |-------------------(Δ bp 137-287)---------------->>
|--------------------------------(Δ bp 87-211)----------------->>
```

FIGURE 3.4

Sequence analysis of mutations targeted *in vivo* within COS cells to the *supFG1* gene in the SV40 vector by the psoralen-conjugated, triple helix-forming oligonucleotide, pso-AG30. Point mutations produced by pso-AG30 and UVA treatment of the vector-containing cells are indicated above each base pair, with the listed base representing a change from the sequence in the top strand. Deletion mutations are presented below the *supFG1* sequence, indicated by the dashed lines. The + signs below the sequence are sites in the *supFG1* gene at which mutations are known to produce a detectable phenotype change, demonstrating that the use of *supF* in this assay does not bias detection of mutations at any particular site. The underlined nucleotides delineate the site targeted for triple helix formation. (From Wang, G. et al., *Mol. Cell Biol.*, 15, 1759, 1995. With permission.)

E. Mutagenesis Induced by High-Affinity Triple Helix Formation

In recent work, we have found that high-affinity triple helix formation, itself, may constitute a DNA lesion that is recognized by repair enzymes and can lead to mutagenesis in mammalian cells.[17] We found that treatment of COS cells containing the pSupFG1 vector with oligonucleotides not conjugated to any mutagen but of high binding affinity led to an induction of mutations in the target *supFG1* gene in the cells at a frequency of 0.27%. The induced mutations, however, were not uniform as in the case of the targeted psoralen adducts. Instead, they were scattered at and around the triplex target site (Figure 3.5), suggesting the occurrence of an error-prone process.

We also found that such high-affinity triple helix formation could stimulate repair synthesis (as measured by radioactive nucleotide incorporation) on a plasmid substrate in HeLa whole-cell extracts. We interpret these findings to suggest that a high-affinity triple helix constitutes an alteration of DNA structure that is recognized by a repair activity and can provoke a mutagenic pathway. The triple helix-induced mutagenesis was not seen in repair-deficient xeroderma pigmentosum cells of complementation group A (XPA), which lack the damage recognition protein in the nucleotide excision repair pathway.[23,24] It was also absent in cells from Cockayne's syndrome, group B (CSB), which are deficient in transcription-coupled repair.[25] When the XPA and CSB cells were transfected with vectors expressing the respective cDNAs of the XPA or CSB genes, the ability of the TFOs to induce mutagenesis in the shuttle vector was restored. These results suggest that the triplex-induced mutagenesis may arise when the triplex formation blocks transcription and thereby induces transcription-coupled repair on an otherwise undamaged template. This process of "gratuitous repair" may lead to mutations at a low frequency. This model is consistent with the recent report that high levels of transcription can generate mutations in yeast, possibly because stalled transcription at natural pause sites may also trigger gratuitous repair.[26]

F. Altered Repair of Targeted Psoralen Adducts in the Context of a Triple Helix

Using oligonucleotides of different lengths, we found that the pattern of mutations produced by triplex-targeted psoralen adducts in the pSupFG1 SV40 shuttle vector in monkey COS cells can be influenced by the associated third strand.[27] Mutations induced by psoralen adducts in the context of a TFO of length 10 (pso-AG10) were the same as those generated by isolated psoralen adducts but were found to be different from those generated in the presence of a TFO of length 30 (pso-AG30) at the same target site. In complementary experiments, HeLa whole-cell extracts were used to assess directly repair of the TFO-directed psoralen adducts *in vitro*. Excision of the damaged DNA was inhibited in the context of the 30-mer TFO but not the 10-mer. These results suggested that an extended triple helix of length 30,

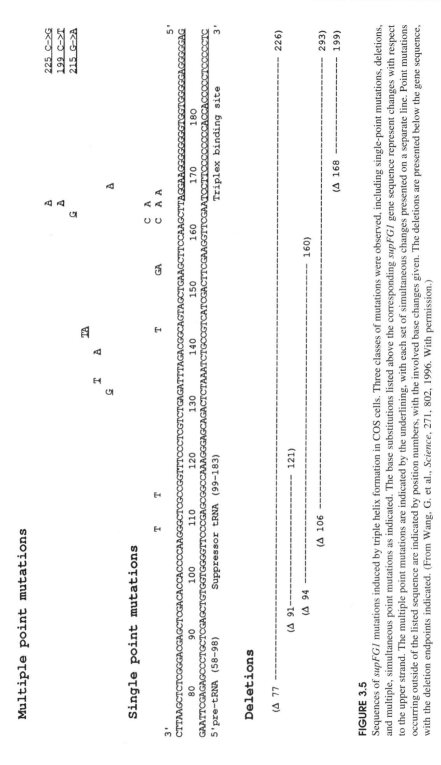

FIGURE 3.5

Sequences of *supFG1* mutations induced by triple helix formation in COS cells. Three classes of mutations were observed, including single-point mutations, deletions, and multiple, simultaneous point mutations as indicated. The base substitutions listed above the corresponding *supFG1* gene sequence represent changes with respect to the upper strand. The multiple point mutations are indicated by the underlining, with each set of simultaneous changes presented on a separate line. Point mutations occurring outside of the listed sequence are indicated by position numbers, with the involved base changes given. The deletions are presented below the gene sequence, with the deletion endpoints indicated. (From Wang, G. et al., *Science*, 271, 802, 1996. With permission.)

which exceeds the typical size of the nucleotide excision repair patch in mammalian cells, can alter repair of an associated psoralen adduct, possibly by blocking the incision steps in repair.

V.   Conclusions

The work reviewed here shows that TFOs conjugated to a mutagenic chemical can mediate targeted mutagenesis of a selected gene both *in vitro* and *in vivo*. By forming triple helices *in vitro*, targeted mutations were induced in λ, SV40, and mouse genomic DNA. Using a TFO with high target site binding affinity, targeted mutagenesis was also achieved in an SV40 viral vector in mammalian cells via intracellular triplex formation. Not only can such TFOs deliver psoralen adducts to the target site, but high affinity third-strand binding itself can be mutagenic, via induction of error-prone repair processes. Interestingly, these findings raise the question of whether intramolecular triple helices, tetraplex DNA, and other possible non-duplex chromosome structures may trigger repair and mutagenesis and may thereby constitute endogenous sources of genomic instability.

Although this work was carried out using an optimized model system, it nonetheless demonstrates the potential use of TFOs for modifying cellular and viral genomes. At present, however, triplex formation is limited to polypurine:polypyrimidine sites. The general applicability of this approach will require advances in nucleotide chemistry so that any site within the genome can serve as a feasible target. In addition, efficient targeting of a chromosomal site in mammalian cells has yet to be demonstrated, and we anticipate that advances in methods for oligonucleotide delivery into cells will eventually be needed in this regard.[28]

Acknowledgments

We thank P. A. Havre, E. J. Gunther, M. M. Seidman, J. George, F. P. Gasparro, L. Narayanan, T. Yeasky, L. Cabral, R. Franklin, and S. J. Baserga for their assistance. This work was supported by the Charles E. Culpeper Foundation, the Leukemia Society of America, the American Cancer Society (CN128), and the NIH (ESO5775 and CA64186).

References

1.   Letai, A. G., Palladino, M. A., Fromm, E., Rizzo, V., and Fresco, J. R., *Biochemistry,* 27, 9108, 1988.

2.   Moser, H. E. and Dervan, P. B., *Science,* 238, 645, 1987.

3.   Francois, J. C., Saison-Behmoaras, T., and Helene, C., *Nucleic Acids Res.,* 16, 11431, 1988.

4.   Maher, L. J., Wold, B., and Dervan P. B., *Science,* 245, 725, 1989.

5.   Postel, E. H., Flint, S. J., Kessler, D. J., and Hogan, M.E., *Proc. Natl. Acad. Sci, USA,* 88, 8227, 1991.

6.   Giovannangeli, C., Thuong, N. T., and Helene, C., *Nucleic Acids Res.,* 20, 4275, 1992.

7.   Birg, F. et al., *Nucleic Acids Res.,* 18, 2901, 1990.

8.   Strobel, S. A., Doucette-Stamm, L. A., Riba, L., Housman, D. E., and Dervan, P. B., *Science,* 254, 1639, 1991.

9.   Perrouault, L. et al., *Nature,* 334, 358, 1990.

10. Havre, P. A. and Glazer, P. M., *J. Virol.,* 67, 7324, 1993.

11. Havre, P. A., Gunther, E. J., Gasparro, F. P., and Glazer, P. M., *Proc. Natl. Acad. Sci. USA,* 90, 7879, 1993.

12. Wang, G., Levy, D.D., Seidman, M. M., and Glazer, P. M., *Mol. Cell. Biol.,* 15, 1759, 1995.

13. Milligan, J. F., Krawczyk, S. H., Wadwani, S., and Matteucci, M. D., *Nucleic Acids Res.,* 21, 327, 1993.

14. Miller, P. S., Bi, G., Kipp, S. A., Fok, V., and DeLong, R. K., *Nucleic Acids Res.,* 24, 730, 1996.

15. Koh, J. S. and Dervan, P. B., *J. Am. Chem. Soc.,* 114, 1470.

16. Griffin, L.C., Kiessling, L. L., Beal, P. A., Gillespie, P., and Dervan, P. B., *J. Am. Chem. Soc.,* 114, 7976, 1992.

17. Wang, G., Seidman, M. M., and Glazer, P. M., *Science,* 271, 802, 1996.

18. Giovannangeli, C., Rougee, M., Garestier, T., Thuong, N. T., and Helene, C., *Proc. Natl. Acad. Sci. USA,* 89, 8631, 1992.

19. Beal, P. A. and Dervan, P. B., *Science,* 251, 1360, 1991.

20. Miller, P. S. and Cushman, C. D., *Bioconjugate Chem.,* 3, 74, 1992.

21. Gasparro, F. P., Havre, P. A., Olack, G. A, Gunther, E. J., and Glazer, P. M., *Nulceic Acids Res.,* 22, 2845, 1994.

22. Glazer, P. M., Sarker, S.N., and Summers, W. C., *Proc. Natl. Acad. Sci. USA,* 91, 5017, 1994.

23. Park, C. H. and Sancar, A., *Proc. Natl. Acad. Sci. USA,* 91, 5017, 1994.

24. Jones, C. J. and Wood, R. D., *Biochemistry,* 32, 12096, 1993.

25. Troelstra, C. et al., *Cell,* 71, 939, 1992.

26. Datta, A. and Jinks-Robertson, S., *Science,* 268, 1616, 1995.

27. Wang, G. and Glazer, P. M., *J. Biol. Chem.,* 270, 22595, 1995.

28. Gewirtz, A. M., Stein, C. A., and Glazer, P. M., *Proc. Natl. Acad. Sci. USA,* 93, 3161, 1996.

Chapter **4**

Protease Footprinting Analysis of Protein-Protein and Protein-DNA Interactions

Roderick Hori and Michael Carey

Contents

I.   Introduction

Proteins form a wide variety of interactions, which enable them to regulate diverse signaling pathways and the assembly of macromolecular complexes. Identification and characterization of these interacting protein surfaces are critical steps toward a mechanistic understanding of how cellular events are regulated. Although the most powerful method for studying such interactions is X-ray crystallography (reviewed in Reference 1), which requires special expertise and equipment, newer methods are available which make identification of interacting surfaces a more tractable goal in a standard molecular biology lab. Here, we describe one such method called protease footprinting, which reveals information on the interaction surfaces and conformational changes of a protein within a macromolecular complex. Facile protein analysis methods will be necessary to study the intricate regulatory networks revealed by the sequencing of the human genome.

In protease footprinting, conceptually similar to DNase I footprinting, partial protease digestion of an end-labeled protein is used to generate a nested set of cleavage products, which is revealed as a ladder on an autoradiograph of an SDS-polyacrylamide gel (Figure 4.1). The surface of the protein that participates in the interaction can be deduced by comparing the cleavage patterns of free and bound protein. Interactions within the complex are manifested either by protected (a "footprint") or enhanced cleavages (Figure 4.1). The protein of interest can be directly end-labeled using a site-specific protein kinase and γ-^{32}P-ATP to phosphorylate a specific site engineered onto the terminus,[2] or indirectly by immunoblotting with an antibody generated against the amino or carboxyl terminus. The position of the interacting residues can be precisely localized by the combination of a protease "sequencing ladder", deletion mutants, and molecular weight markers. The sequencing ladder is generated by denaturing the end-labeled protein and performing limited digestion with site-specific proteolytic reagents such as endoproteinase Lys-C or cyanogen bromide (CNBr).

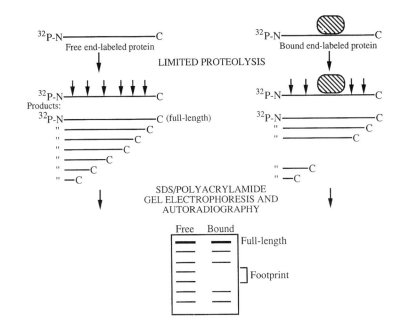

FIGURE 4.1

Protease footprinting scheme: Free and bound end-labeled protein is subjected to limited proteolysis generating a nested set of cleavage products. In the case of the bound protein, the ligand (striped oval) protects specific sites from proteolysis, resulting in the absence of those digestion products. The cleavage products are electrophoresed on a denaturing polyacrylamide gel, and the digestion pattern is visualized by autoradiography or a phosphorimager. The interaction is observed as a lack of the protected cleavage products or a "footprint". The position of these protected cleavage products can be mapped by a variety of methods (see Section V).

A. What Information Can Be Obtained from Protease Footprinting?

Protease footprinting has been successfully employed to probe interactions within a wide variety of macromolecular complexes. The conformational change that occurs when yeast topoisomerase II binds ATP[3] and the ability of a point mutation to prevent the communication of this change between subunits[4] were two of the original applications of the technique. The method also reveals the interaction between proteins and small ligands, as illustrated by the mapping of apomyoglobin's binding site for heme.[5]

Protease footprinting has most frequently been used to identify the surfaces of a protein that interact with DNA or RNA as in the cases of TFIIIA, the cAMP receptor protein (CRP), the endonucleases I-PorI and I-DmoI, and Rev.[6-9] The method has been useful in examining protein-protein interactions. Examples include the contacts among the *Escherichia coli* RNA polymerase subunits and the interaction of the herpes virus transactivator VP16 and the co-activator TAF40 with the RNA polymerase II (pol II) general transcription factor TFIIB.[10-12] Further, the method proved particularly powerful for studying both protein-protein and protein-DNA interactions of the pol II factor TFIIA within a ternary complex containing TFIIA, the TATA-box binding protein (TBP), and promoter DNA.[13] In the cases of the TFIIA-containing ternary complex and the CRP-DNA complex, the interacting surfaces deduced by protease footprinting were in excellent agreement with the crystal structures.[7,13] In many of the aforementioned examples, the interaction sites could be mapped to 10- to 20-amino acid regions within the protein. Given the variety of interactions successfully probed by the method, it should be generalizable to most situations.

B.   Historical Perspective

The use of proteolysis to probe protein structure is not a new concept. The domain structure of a protein is traditionally surmised by proteolysis of a protein followed by peptide sequencing or mass spectrometry of the resistant products. Proteolytic-resistant regions have often proven to define domains within a protein, while proteolytic-sensitive sites generally define interdomain spacers. One recent example was the analysis of the zinc finger DNA-binding domains of TFIIIA. The fingers were initially discovered as 3-kDa protease-resistant fragments,[14] which, after comparison with the amino acid sequence, revealed the existence of small repeating units, each now known to encode a zinc-nucleated DNA binding region. Proteolysis studies of the bacteriophage lambda repressor defined an amino-terminal domain, which contained the DNA-binding domain, and a carboxyl-terminal domain, which mediated dimerization.[15] Peptide sequencing of the fragments precisely defined each domain, and thermal denaturation profiles confirmed the presence of two domains in the intact protein that possessed melting temperatures equivalent to those of the proteolytic fragments. Proteolysis can be used to study protein-ligand interactions by comparing, on a Coomassie-stained gel, the cleavage products obtained in the presence and absence of a ligand. However, direct sequencing of each product is necessary to identify the position of the cleavages and to determine which ones are influenced by ligand.

The major advances that have made these experiments more amenable to a standard biology laboratory are direct ^{32}P end-labeling of the protein, high-resolution gel systems to fractionate the cleavage products, and proteolytic sequencing ladders to map the footprints. The high-resolution gel system and sequencing ladders permit accurate identification of cleavage products without resorting to mass spectrometry or peptide sequencing, while the ^{32}P end-label provides a nested set which greatly facilitates site mapping and imparts higher sensitivity and ease in quantitation, particularly with the increased availability of phosphorimagers. Finally, like the evolution of the DNase I footprinting technique, modified procedures that produce higher resolution cleavage maps have been or are currently being developed. These include the use of hydroxyl radical as a cleavage reagent[7,11,12] and the combination of small chemical modification reagents which influence proteolysis[16] (for more details, see Section VII.A).

II.   End-Labeling the Protein

A.   Direct End-Label

The catalytic subunit of protein kinase A (PKA) and g-^{32}P-ATP can be used to label proteins directly. The catalytic subunit can be obtained from Sigma (P-2645) under the name bovine heart muscle kinase (HMK), and it phosphorylates a specific recognition sequence termed the HMK tag (Arg-Arg-Ala-Ser-Val/Leu; see Reference 2). A specific end-label is generated by engineering the recognition site onto one of the termini of the protein; the remainder of the protein must lack endogenous kinase sites. Several proteins have been ^{32}P-labeled in this fashion, including myoglobin, E. coli RNA polymerase subunits, TFIIB, the subunits of TFIIA, endonucleases, and Rev.[5,8-11,13] If there are any endogenous sites, they must be eliminated or indirect end-labeling will have to be employed (see Section II.B). It is imperative to exclude the possibility that the added tag or its phosphorylation influence the activity of the protein.

The terminal HMK tag can be engineered by several methods, including oligonucleotide-directed mutagenesis[5] or by cloning the cDNA encoding the protein into an expression vector containing an in-frame HMK tag.[10] Typically, the HMK-tagged protein can be purified by the same methods used to purify the untagged version. It is very important that the purified protein be absolutely intact because any degraded products containing the HMK tag will also be phosphorylated and generate background signal, thereby confounding the analysis. To circumvent this problem, a FLAG™ or His$_6$-tag can be fused to the terminus opposite the HMK tag and used to purify the protein in a single-step procedure using immunoaffinity or nickel affinity chromatography resins (Kodak; Qiagen; Invitrogen), i.e., as performed for myoglobin and TFIIA.[5,13] Because the tags employed for purification and labeling are at opposite termini, only full-length molecules should in principle become end-labeled.

To determine the specificity of end-labeling, equal amounts of the HMK-tagged and the wild-type proteins should be compared for their ability to incorporate ^{32}P in the presence of HMK and γ-^{32}P-ATP. A small portion of the reaction products are fractionated on a sodium dodecyl sulfate (SDS)-polyacrylaminde gel and the gel is autoradiographed to determine the incorporation efficiency (Figure 4.2). If significant amounts of labeling occur with the wild-type protein, the amount of kinase and the time of incubation should be optimized to minimize the background.

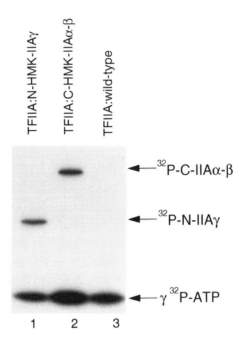

Labels on lanes: TFIIA:N-HMK-IIAγ, TFIIA:C-HMK-IIAα-β, TFIIA:wild-type

$\longleftarrow$ ^{32}P-C-IIAα-β

$\longleftarrow$ ^{32}P-N-IIAγ

$\longleftarrow$ γ ^{32}P-ATP

1   2   3

FIGURE 4.2
Specific end-labeling of HMK-tagged TFIIA. 2 μg each of the RNA polymerase II general transcription factor TFIIA in which the γ (small) subunit (lane 1), α–β (large) subunit (lane 2), or neither subunit (lane 3) contains an HMK tag were incubated with HMK and γ-^{32}P-ATP. A small portion of each reaction was electrophoresed on a 15% SDS-polyacrylamide gel and autoradiographed.

1.   Experimental procedure for end-labeling a protein

The precise reaction conditions will depend on several factors. A time course of the reaction should be performed to optimize the signal-to-background ratio. The amount of protein and γ-^{32}P-ATP required will rely on the efficiency of the ^{32}P incorporation and the amounts of end-labeled protein required to obtain formation of specific complexes (see Section III). HMK is resuspended in 6 mg/ml DTT, as specified by Sigma, at a concentration of 5 U/μl, frozen on dry ice in small aliquots (for one use), and stored at –70°C. Using these conditions, we have observed minimal loss of HMK activity (unpublished observations). The protocol below lists the reaction conditions for the phosphorylation of TFIIA shown in Figure 4.2. The 100-μl reaction mixes contain 20 μl 5× HMK buffer (100 mM Tris, pH 7.6; 500 mM NaCl; and 60 mM MgCl$_2$; see Reference 2), 6 μl HMK (30 U), 4 μl γ-^{32}P-ATP (150 mCi/ ml; specific activity: 6000 Ci/mmol; Amersham PB15068), and 2 μg of wild-type or HMK-tagged TFIIA. After 45 min at 25°C, the reactions are quenched by the addition of 1 μl of 100 mM ATP and incubation on ice. The protein is divided into aliquots, frozen on dry ice, and stored at –70°C. Another variation of this procedure is to use γ-^{33}P-ATP rather than γ-^{32}P-ATP,[11] which yields a sharper signal for quantitation, produces less radiolysis, and has a longer half-life (R. Ebright, personal communication), but does produce a weaker signal. If, after comparing HMK-tagged and untagged protein, the signal-to-background ratio is low, replacing Tris with HEPES increases the specificity in some instances (M. Haykinson and R. Johnson, personal communication).

B.   Indirect End-Label

A protein can be indirectly end-labeled by performing an immunoblot using a primary antibody (either mono- or polyclonal), generated against the terminal amino acids of the protein. The binding of the primary antibody acts as an indirect end-label, which can be detected with a secondary antibody and a chemical or enzymological staining system. This approach generates the same nested set of digestion products that would be produced with a direct ^{32}P end-label. This strategy has been used in studies that map the DNA-binding site of CRP[7] and the interaction of the σ subunit with the other subunits of *E. coli* RNA polymerase.[12]

Drawbacks to this method are the time and cost required to generate antibodies. An alternative is to engineer a so-called "epitope-tag" onto one terminus of the protein.[17] An epitope-tag was used in a study to investigate the conformational changes made by topoisomerase II upon binding ATP.[3,4] The advantage of an epitope-tag is the commercial availability of high affinity antibodies such as ones to FLAG™ (Kodak), hemagglutinin (HA; Boehringer), and c-*myc* epitopes (Boehringer). In the case of the HA epitope, a primary antibody conjugated to multiple molecules of horseradish peroxidase (HRP) is commercially available, permitting the use of sensitive chemiluminescence detection methods, without the need for a secondary antibody.

Direct labeling offers increased sensitivity, allows faster detection because the time and physical manipulation necessary to perform an immunoblot are not required, and eliminates the need to produce antibodies. Indirect end-labeling allows the use of the completely wild-type protein (an advantage that is lost if one decides to use an epitope tag) and eliminates the need to use radioactivity. For simplicity, the molecule to be labeled directly or indirectly will be referred to generically as "end-labeled" in the rest of this chapter.

III.   Complex Formation

A.   Assembly of Protein-Protein Complexes

Assembly of the end-labeled protein into a complex must be quantitative. The presence of unbound end-labeled protein during proteolysis will prevent the detection of a footprint against the background cleavage signal. Several strategies have been used to generate such stoichiometric complexes. When the interaction is strong, it may be possible to obtain complete saturation of the end-labeled protein with its partner without

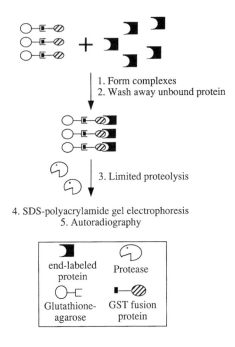

FIGURE 4.3

Schematic of protease footprinting using an affinity resin. The end-labeled protein is incubated with the affinity resin and washed extensively to isolate complexes, which are then subjected to limited proteolysis. The cleavage products are electrophoresed on denaturing gels, and autoradiography is used to visualize the digestion pattern.

further purification, as in the case of the interaction of myoglobin with HAL-19, a high-affinity monoclonal antibody,[5] or in the case of ATP binding to topoisomerase II.[4]

When the affinities are too low or if for other reasons one is unable to saturate the binding of the end-labeled protein in solution, alternative schemes must be employed to isolate the complexes. One method is to purify the desired complexes away from their separate components by chromatography; for example, *E. coli* core RNA polymerase and holoenzyme (core + σ) were purified as separate complexes from *E. coli* to identify the binding sites of the σ subunit within the core.[12] Similarly, myoglobin and its apo-form were overexpressed in *E. coli* and purified separately to analyze the interaction of the heme co-factor.[5] Finally, affinity resins provide an immobilized support that has been used to purify complexes simply and quickly,[10] as described below.

1.   Affinity resins

Affinity resins provide a means to rapidly isolate protein-protein complexes, if the desired complexes cannot be assembled and purified by other methods. One common affinity resin is glutathione-S-transferase (GST) fusion proteins bound to glutathione-agarose. To assemble complexes, the end-labeled protein is incubated with the resin displaying the GST-fusion protein, and bound protein is purified after extensive washing (Figure 4.3). For example, to form complexes between TFIIB and VP16, ^{32}P-labeled TFIIB was bound to a GST-GAL4-VP16 affinity resin, and unbound TFIIB was removed by washing. To determine the specificity of the binding, the end-labeled protein is also incubated in parallel with equal amounts of a resin displaying GST alone. A small portion of each bound fraction is then analyzed by SDS-polyacrylamide gel electrophoresis and autoradiography to determine the amounts bound to each resin. In order to perform protease footprinting, the ratio of specific to nonspecific binding should be at least 5:1 (i.e., 80% of the end-labeled protein is retained via a specific interaction). After establishing binding conditions, the amount of disassociation is minimized in the actual proteolysis experiments by performing the digestions immediately after complex purification.

Procedure for generating the GST-fusion protein affinity resin

GST-fusion proteins can be constructed using the pGEX vectors (Pharmacia). The pGEX vectors contain the hybrid tac promoter of *E. coli*, driving expression of the gene encoding GST. Positioned immediately downstream from the GST coding sequence are restriction sites for generating in-frame fusions between GST and the cDNA encoding the "target" protein. pGEX vectors also carry the *lacI*q allele (which overexpresses Lac repressor), allowing them to be grown in most common *E. coli* hosts. Bacteria transformed with the expression vector are grown at 37°C to an A_{600} of 0.6, and expression is induced by the addition of isopropyl-1-thio-β-D-galactopyranoside to 0.4 mM. After shaking at 37°C for 1.5 hr, the cells are collected by centrifugation at 4000 rpm in a Sorvall GSA rotor, washed with buffer A (20 mM HEPES, pH 7.9; 2 mM EDTA) containing 0.4 M KCl and resuspended in buffer A containing 0.4 M KCl, 1 μg/ml pepstatin A, 1 μg/ml leupeptin, 50 μg/ml benzamidine, 0.5 mM phenylmethylsulfonyl fluoride (PMSF) and 0.5 mM DTT. All subsequent steps are performed at 4°C. The cells are lysed by sonication, the cell lysate is separated from the insoluble fraction by centrifugation at 10,000 rpm in a GSA rotor, and the lysates are incubated with the glutathione-agarose affinity matrix (Sigma; Pharmacia) on a nutator (Clay Adams). The resin is collected either by gentle centrifugation or on a scintered glass funnel, then washed extensively with buffer A containing 0.4 M KCl, 0.5 mM PMSF, and 0.5 mM DTT.

Within an experiment, the amount of protein bound to each affinity resin must be identical and can be normalized by measuring the protein concentration by the method of Bradford[18] and the integrity of protein bound to each affinity resin by electrophoresing a portion on a SDS-polyacrylamide gel followed by staining with Coomassie blue. Since GST-fusion protein preparations often contain some wild-type GST along with the fusion protein, it is important to measure both factors. The presence of wild-type GST is presumably due to proteolysis, which occurs during purification, at the junction of the fusion protein.

Alternative affinity resins can be generated by constructing fusions between the His$_6$-tag (Invitrogen and Qiagen) or immuno-tags, such as the FLAG™ epitope (Kodak). After expressing these fusion proteins in bacteria, they can be bound to either nickel- or antibody-affinity resins.

2. Considerations for protein-protein complex formation using an affinity resin

The reaction conditions should maximize the specific assembly between the end-labeled protein and ligand. This is accomplished by altering the amount of input end-labeled protein and the buffer used during the reaction. Ideally, the buffer would be similar to those used in other *in vitro* assays. For example, assembly of complexes containing a transcriptional activator and a general transcription factor are performed under conditions similar to those used during *in vitro* transcription assays.[10] Additional BSA and low concentrations of detergent (e.g., 0.05% NP-40) are often added to minimize nonspecific interactions.

In a typical complex assembly reaction, 10 μg of the immobilized GST-fusion protein (approximately 0.2 nmol) or GST alone are equilibrated for 5 min at room temperature in 200 μl binding buffer containing 200 μg/ml BSA to absorb any nonspecific binding sites. The resin is collected by centrifugation at 4500 rpm for 2 min at room temperature in a microcentrifuge. This equilibration step is repeated and then 1 ml of reaction buffer containing end-labeled protein is added. (The optimal amount of input end-labeled protein can be determined by varying the amount added in each reaction.) The radioactivity in each reaction is determined using a Geiger counter. After incubating for 30 min at 25°C on a nutator, the bound fraction is collected by centrifugation at 4500 rpm in a microcentrifuge for 2 min at 25°C. The amount of labeled protein remaining in each reaction is measured. One ml of reaction buffer is added and the samples are nutated for 30 sec to wash the resin, then the bound fraction is collected by centrifugation as before. The wash step is repeated until only residual protein remains bound to the GST resin or washing does not remove additional end-labeled protein. Usually a total of 3 washes is sufficient. An equal volume of 2× SDS-PAGE sample buffer (100 mM Tris, pH 6.8; 200 mM DTT; 4% SDS; 0.2% bromophenol blue; 20% glycerol) is added, and the samples are heated at 95°C for 4 min. The samples are centrifuged at 14,000 rpm in a microcentrifuge to pellet any protein that is irreversibly bound to the resin. A portion of each bound fraction is electrophoresed on an SDS-polyacrylamide gel and autoradiographed, and the ratio of specific binding is determined. If desired, the gel can be stained with Coomassie blue prior to autoradiography to ensure that equal amounts of GST and GST-fusion protein were present in each binding reaction. If the specificity is too low for protease footprinting, improvements can be made by varying the amount of input end-labeled protein, salt concentrations, BSA, or detergent or by using different detergents. To perform protease footprinting after complex assembly, the affinity resin displaying GST alone should be washed less extensively so it retains approximately the same amount of end-labeled protein as the specific affinity resin.[10]

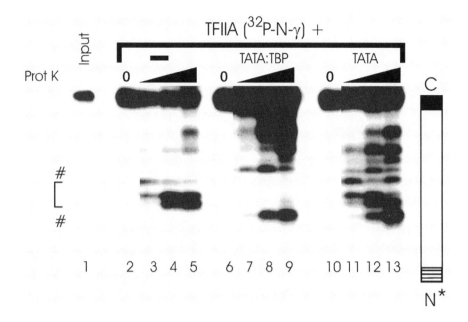

FIGURE 4.4

Protease footprint of the TFIIA:TBP:TATA box ternary complex. Ternary complexes containing TFIIA end-labeled on the amino terminus of the γ subunit, TBP, and a TATA box oligonucleotide were assembled, purified using a TATA-box affinity resin, and digested with increasing amounts of proteinase K (lanes 7 to 9) in parallel with free TFIIA (lanes 3 to 5) and TFIIA incubated in the presence of the TATA box (lanes 11 to 13). Mock digestions (0) are shown in lanes 2, 6, and 10. Lane 1 contains the input TFIIA used in the complex assembly reactions. The region of the footprint is indicated by a bracket on the left, and the pound signs denote sites that are hypersensitive in the presence of the TATA-box oligonucleotide. A schematic of the γ subunit of TFIIA is shown on the right, where the carboxyl terminus, amino terminus, end-label (asterisk), HMK tag (striped region), and His_6 tag (black region) are denoted.

B.   Protein-DNA Complexes

In some cases involving DNA-protein complexes, DNA affinity resins will be required to isolate complexes. The considerations for assembling protein-DNA complexes on an affinity resin are similar to those discussed in Section III.A.2. The affinity resin can be generated by coupling the DNA fragment, using a biotin moiety, to a streptavidin matrix (Dynal or BRL). The biotin is introduced by using the Klenow fragment and biotinylated nucleoside triphosphates to repair a 5' over-hang of a DNA restriction endonuclease or during synthesis, if a synthetic oligo-nucleotide is used. In a study which characterized the interactions made by TFIIA within the TFIIA:TBP:TATA box ternary complex,[13] the complexes were isolated using a TATA box oligonucleotide affinity resin; a typical protease footprinting experiment is shown in Figure 4.4.

TABLE 4.1
Type and Specificity of Proteases Used in Footprinting

Protease	Type	Cleavage Specificity
Alkaline protease	Serine	Neutral and aromatic
Chymotrypsin	Serine	Aromatic
Clostripain	Cysteine	Arginine
Elastase	Serine	Small, uncharged
Endoproteinase Arg-C	Serine	Arginine
Endoproteinase Asp-N	Metallo	Aspartic acid and cysteine
Endoproteinase Glu-C	Serine	Glutamic acid and aspartic acid
Endoproteinase Lys-C	Serine	Lysine
Pronase	Mixture	Nonspecific
Proteinase K	Serine	Nonspecific, but prefers aromatic or hydrophobic residues
Thermolysin	Metallo	Hydrophobic
Trypsin	Serine	Arginine and lysine

IV. Proteolysis

A. Enzymatic Proteases

Broad-specificity enzymatic proteases are generally more useful for identifying protein interactions because they access a larger portion of the surface of a protein. Proteases which have been used in various footprinting studies include elastase, alkaline protease (subtilisin), pronase, thermolysin, and proteinase K, although proteases with more limited specificity such as chymotrypsin, trypsin, endoproteinase Asp-N, endoproteinase Arg-C, endoproteinase Lys-C, and endoproteinase Glu-C have also been employed.[3-6,8-10,13] Table 4.1 summarizes some general properties of these proteases.

The length of digestion is crucial because any dissociation of labeled protein during the proteolysis reaction will lead to a background cleavage ladder on the resulting gel. Shorter incubation times decrease the chances for both background cleavage and secondary cleavage events in which the cleaved protein is recleaved, an event that might confound the identification of true interacting residues. Typically, proteases were incubated with the complex for 10 or 15 min when examining stable states such as heme binding by myoglobin or allosteric transitions in topoisomerase II.[3-5] Incubations of 10 or 15 min were also used to assay protein-protein and protein-DNA complexes having a high affinity or that could be trapped in a stable state such as the binding of an antibody to myoglobin and of endonucleases to DNA.[5,9] Shorter incubations of 1 or 1.5 min were used for less stable complexes such as TFIIIA binding to its DNA site, formation of a protein-DNA complex containing TFIIA, and protein-protein complexes containing other pol II transcription factors.[6,10,13] The amount of protease should be adjusted so that 20 to 40% of the full-length protein is cleaved during the digestion. Protease footprinting experiments have been performed using either a range of protease amounts [6,8,10,13] or by using one optimal concentration.[3-5,9]

The amounts of protease required to obtain the optimal proteolysis pattern can be determined by titrating each protease through a wide range and examining the resulting cleavage pattern. All proteases used should be stored in single-use aliquots at −70°C to ensure consistency in performance. We have found that after one freeze/thaw cycle, many proteases are less active presumably due to either denaturation during the freeze/thaw cycle or autoproteolysis (unpublished observations). The proteases are ideally stored at a stock concentration of 1 mg/ml. The protease levels are optimized by serial dilution into mixtures containing the end-labeled protein. After a short time period (1 to 15 min), the reaction is terminated by addition of an appropriate protease inhibitor (e.g., PMSF) and freezing the sample on dry ice. When all the reactions are terminated, an equal volume of 2× SDS-PAGE loading dye is added and the samples are heated at 95°C for 4 min. A portion of each reaction is analyzed on a SDS-polyacrylamide gel and autoradiographed to visualize the extent of proteolysis.

B.   Chemical Proteases

Fe-EDTA complexes, in the presence of H_2O_2, have been used instead of enzymatic proteases to map protein interactions.[7,11,12] Although the mechanism of proteolysis is still not fully understood, these reagents have also been employed to map DNA interactions, and the chemistry involved in DNA cleavage is well characterized. In the nuclease reaction, an electron from Fe(II) reacts with H_2O_2 to generate a hydroxide ion and a hydroxyl radical, the latter of which cleaves DNA by abstracting a hydrogen atom from the deoxyribose moiety (reviewed in Reference 19). It has been suggested that protein scission occurs through oxidative cleavage via a diffusible hydroxy radical species,[20] although a hydrolytic mechanism has also been proposed.[21] The advantage of this methodology is the ability to probe a much larger proportion of the protein because of the nonspecific nature of digestion and the smaller size of hydroxyl radicals relative to protein proteases.

Fe-EDTA generated from $(NH_4)_2Fe(II)(SO_4)_2$[7,11] or Fe(III)-EDTA[12] have been used in protease footprinting studies. Glycerol and thiols quench the cleavage reaction and must be eliminated from the buffers or greatly minimized during proteolysis.[12] These reagents can be removed by using either buffers that do not contain them during the last stage of the purification[11] or a small spin column to desalt the mixture immediately prior to digestion.[12]

The contacts made by the α subunit during the assembly of *E. coli* RNA polymerase were examined using Fe-EDTA footprinting.[11] *E. coli* RNA polymerase (or subcomplexes) containing an end-labeled α subunit at 4 μ*M* were incubated in

10-μl reactions containing 10 mM MOPS-NaOH (pH 7.2), 5 mM Tris-HCl, 125 mM NaCl, 2 μM ZnCl$_2$, 1% glycerol, 2 mM EDTA, 1 mM (NH$_4$)$_2$Fe(II)(SO$_4$)$_2$, 1 mM H$_2$O$_2$, and 20 mM sodium ascorbate. The reactions were initiated by the addition of freshly prepared (NH$_4$)$_2$Fe(II)(SO$_4$)$_2$ and EDTA and terminated after 30 min by the addition of 1/2 volume of 3× sample buffer (150 mM Tris-HCl, pH 7.9; 36% glycerol; 12% SDS; 6% β-mercaptoethanol; 0.01% bromophenol blue), which completely terminates the reaction. The samples were then fractionated by electrophoresis on tricine-SDS-polyacrylamide gels (see Section VI).[22] The interactions of the σ subunit within *E. coli* RNA polymerase were studied using preformed Fe(III)-EDTA[12] rather than (NH$_4$)$_2$Fe(II)(SO$_4$)$_2$, slightly different concentrations of sodium ascorbate and H$_2$O$_2$, and only 1-min reaction times. Although both studies are excellent starting points, they illustrate the need to optimize the amounts of each reagent and the time of digestion.

V.   Assigning the Positions of the Cleavages

The position of the cleavage sites can be assigned by using a combination of four approaches — "sequencing" ladders,[6,10,11,13] deletion mutants, prestained molecular weight standards (BioRad), and the known amino acid sequence. The pattern generated by sequencing proteases provides landmarks along the cleavage ladder, while the other reagents are used to identify the positions of those landmarks.

In a manner analogous to Maxam-Gilbert sequencing ladders in DNase footprinting, a ladder defining specific amino acid residues is electrophoresed alongside the experimental lanes to map the borders of the footprint. To generate the sequencing ladder, the end-labeled protein is denatured prior to partial digestion to increase the accessibility of all or nearly all of the potential cleavage sites. However, not all sites are cleaved equally, presumably due to either preferences of the protease for neighboring residues or some residual structural features being maintained after denaturation, preventing access of the protease. Because of the end-label, a nested set of cleavage products will be generated. Sequencing proteases include endoproteinase Asp-N, endoproteinase Lys-C, clostripain, and endoproteinase Glu-C, which cleave on the amino-terminal side of aspartic acid or the carboxyl-terminal side of lysine, arginine, and glutamic/aspartic acid, respectively;[6,10,11,13] or chemicals which cleave particular residues such as CNBr.[6,11]

Cleavage at methionine residues can be obtained by using CNBr. In one example, the end-labeled α subunit from RNA polymerase was incubated for 20 min at 25°C in a buffer adjusted to pH 2 and containing 500 mM CNBr and 0.4% SDS for 20 min and terminated by lyophilization.[11] Arginine-specific sequencing ladders have been obtained by denaturing protein at 65°C for 30 min in a buffer containing 20 mM HEPES, pH 7.9; 100 mM KCl; 2 M urea; and 0.2% SDS. The reaction mixture was transferred to 37°C, and 2 μg of clostripain (Promega) were added. Aliquots were removed at 2, 6, 18, and 54 min, and the reaction was terminated by adding EDTA to 50 mM and 1 μl of 10 mg/ml BSA along with freezing the samples on dry ice.[13] After the digestion, an equal volume of 2× SDS-PAGE loading dye was added, the samples were heated at 95°C for 4 min, and a small aliquot of each time-point was electrophoresed on a denaturing gel. In each case, the cleavage pattern was visualized by either autoradiography or a phosphorimager.

Since the cleavage reagents are chosen because they have a restricted specificity, the predicted digestion pattern based on the known amino acid sequence can be compared with the observed cleavage pattern. Still, not all the sites are digested or digested at the same rate (see above), which can leave some ambiguity in the assignments. To help assign the cleavage sites, prestained molecular weight standards are used to provide a rough estimate of the molecular mass of each cleavage fragment. Finally, the precise position of particular amino acids can be determined by making deletion mutants at known positions. This can be accomplished by constructing deletions from the terminus opposite the end-label by using unique restriction sites, or PCR. The resulting truncated proteins can usually be expressed in *E.coli* and purified in a manner similar to the wild-type protein. This approach provides markers that define the exact position of that residue and act as checks for the register assigned to the sequencing ladder.

VI.   Gel Systems

The gel system and the dimensions of the gels used will depend upon the resolution required to visualize the interactions of interest. Analogous to DNA sequencing, gels that are longer and thinner will provide better resolution and, if autoradiography is used, a sharper signal. Both Tris-tricine-SDS (tricine)[22] and Tris-glycine-SDS (glycine) polyacrylamide gels[23] have been used in protease footprinting experiments.[3-13] The advantage of tricine gels lies in their ability to resolve polypeptide chains down to 1-kDa and higher resolution of small polypeptide chains. A more detailed discussion of tricine gels is available in Schagger and von Jagow.[22] The choice of gel system will depend primarily on the distance from the end-label where the interaction occurs and the resolution required to separate adjacent cleavage products.

VII.   Specialized Cases

A.   Use of Reversible and Irreversible Modification Reagents

The inability of enzymatic proteases to cleave large portions of a protein due to its structural integrity can be circumvented by the use of small reversible and irreversible modification reagents.[16] Such reagents also can give a more detailed picture of an interaction due to their small size. The regions of vaccinia virus topoisomerase that interact with DNA were identified by lightly treating complexes and free protein with citraconic anhydride, which reversibly modifies lysine residues. After quenching the reaction, the protein was denatured and incubated with an excess of N-hydroxysuccinimide acetate, an irreversible lysine modification reagent. After quenching this reaction, the citraconylated lysines were deblocked and the protein was digested with endoproteinase Lys-C, which cleaves at the unmodified lysine residues (i.e., those that were initially citraconylated). The interaction of topoisomerase with DNA protected particular sites from citraconylation, and, upon subsequent modification with hydroxysuccinimide, those sites were digested at a much lower frequency than in the free protein. The digestion products were separated on a denaturing gel and indirectly end-labeled to visualize the cleavage sites protected by DNA.

B.   Covalently Attached Proteolytic Reagents

One goal of studying protein structure is to determine which regions of protein are adjacent in the three-dimensional structure of a protein or complex. The parts of a protein that are adjacent to a particular residue can be examined by using a proteolytic reagent attached covalently to that position. In a study of the lactose permease, the protein was engineered to contain unique cysteine residues, which were derivatized with 1,10-phenanthroline.[24] Proteolysis was initiated by the addition of copper sulfate and ascorbate (and sometimes H_2O_2), terminated by the addition of neocuproine, separated on denaturing gels, and immunoblotted with an antibody specific to the carboxyl terminus to generate an indirect end-label. The results of this study identified which of the 12 transmembrane helical domains were adjacent to helix V. Alternatively, Fe-EDTA has been exploited as a proteolytic reagent when derivatized to proteins and used to distinguish which regions of cytochrome bd are adjacent to its quinone-binding domain[25] and to characterize the native structure and conformational state during unfolding of staphylococcal nucleases.[26]

C.   Combining Mass Spectrometry
    and Proteolysis

The structure of a protein and its interaction with a ligand can also be probed by performing proteolysis of the wild-type protein under various conditions and using matrix-assisted laser desorption/ionization (MALDI) mass spectrometry,[27] rather than the methods discussed in Section V, to determine the sites of proteolysis. MALDI mass spectrometry is well suited for protease footprinting studies because it can measure the molecular masses of complex mixtures of peptide fragments, produces accurate mass determinations to determine the sites of digestion, is tolerant to many buffer conditions, and can be performed in minutes.[27] This methodology was employed to probe the solution structure of the transcription factor Max, both free and bound to DNA, and the results were compared to the crystal structure. Consistent with the crystallographic data, the proteolysis study indicates the basic region in the amino-terminus of Max bound to its specific DNA site. In addition, a comparison of the proteolysis patterns in the absence and presence of DNA provides evidence for the "induced fit" model in which the amino-terminus of Max undergoes a transition from a random coil to helical domains in the presence of its cognate recognition site.[27]

VIII. Summary

Protease footprinting provides a straightforward biochemical method for identifying protein interactions and conformational changes within a complex. As more proteins that regulate biological functions are identified, the size and number of assemblies being studied increases and the complexity of the interactions expands, generating great potential for protease footprinting to elucidate structural features that cannot be obtained by other standard methods such as crystallography. The development of small chemical cleavage reagents will be of particular use for these types of studies.

Acknowledgments

We would like to thank Richard Ebright, Michelle Martin, Tian Chi, Dean Tantin, and David Sigman for their excellent input into the writing of this manuscript. This study was supported by an NIH grant to MC (GM46424) and an NIH postdoctoral fellowship and a Jonsson Cancer Center fellowship to RH.

References

1. Sjolin, L., *Drug Des. Discov.,* 9, 261, 1993.

2. Li, B. L., Langer, J. A., Schwartz, B., and Pestka, S., *Proc. Natl. Acad. Sci. USA,* 86, 558, 1989.

3. Lindsley, J. E. and Wang, J. C., *Proc. Natl. Acad. Sci. USA,* 88, 10485, 1991.

4. Lindsley, J. E. and Wang, J. C., *Nature,* 361, 749, 1993.

5. Zhong, M., Lin, L., and Kallenbach, N. R., *Proc. Natl. Acad. Sci. USA,* 92, 2111, 1995.

6. Bogenhagen, D. F., *Mol. Cell. Biol.,* 13, 5149, 1993.

7. Heyduk, E. and Heyduk, T., *Biochemistry,* 33, 9643, 1994.

8. Jensen, T. H., Leffers, H., and Kjems, J., *J. Biol. Chem.,* 270, 13777, 1995.

9. Lykke-Anderson, J., Garret, R. A., and Kjems, J., *Nucl. Acids Res.,* 24, 3982, 1996.

10. Hori, R., Pyo, S., and Carey, M., *Proc. Natl. Acad. Sci. USA,* 92, 6047, 1995.

11. Heyduk, T., Heyduck, E., Severinov, K., Tang, H., and Ebright, R. H., *Proc. Natl. Acad. Sci. USA,* 93, 10162, 1996.

12. Greiner, D. P., Hughes, K. A., Gunasekera, A. H., and Meares, C. F., *Proc. Natl. Acad. Sci. USA,* 93, 71, 1996.

13. Hori, R. and Carey, M., *J. Biol. Chem.,* 272, 1180, 1997.

14. Miller, J., McLachlan, A. D., and Klug, A., *EMBO J.,* 4, 1609, 1985.

15. Pabo, C. O., Sauer, R. T., Sturtevant, J. M., and Ptashne, M., *Proc. Natl. Acad. Sci. USA,* 76, 1608, 1979.

16. Hanai, R. and Wang, J. C., *Proc. Natl. Acad. Sci. USA,* 91, 11904, 1994.

17. Matsudaira, P., Jakes, R., Cameron, L., and Atherton, E., *Proc. Natl. Acad. Sci. USA,* 82, 6788, 1985.

18. Bradford, M. M., *Anal. Biochem.,* 72, 248, 1976.

19. Price, M. A. and Tullius, T. D., *Methods Enzymol.,* 212, 194, 1992.

20. Platis, I. E., Ermácora, M. R., and Fox, R. O., *Biochemistry,* 32, 12761, 1993.

21. Rana, T. M. and Meares, C. F., *Proc. Natl. Acad. Sci. USA,* 88, 10578, 1991.

22. Schagger, H. and von Jagow, G., *Anal. Biochem.,* 166, 368, 1987.

23. Laemmli, U. K., *Nature,* 227, 680, 1970.

24. Wu, J., Perrin, D. M., Sigman, D. S., and Kaback, H. R., *Proc. Natl. Acad. Sci. USA,* 92, 9186, 1995.

25. Ghaim, J. B., Greiner, D. P., Meares, C. F., and Gennis, R. B., *Biochemistry,* 34, 11311, 1995.

26. Ermacora, M. R., Ledman, D. W., Hellinga, H. W., Hsu, G. W., and Fox, R. O., *Biochemistry,* 33, 13625, 1994.

27. Cohen, S. L., Ferre-D'Amare, A. R., Burley, S. K., and Chait, B. T., *Protein Sci.,* 4, 1088, 1995.

Gene Expression

Chapter 5

Studying Gene Expression on Tissue Sections Using *In Situ* Hybridization

Urs Albrecht, Hui-Chen Lu, Jean-Pierre Revelli,
Xiao-Chun Xu, Reuben Lotan, and Gregor Eichele

Contents

I.   Introduction

The identification of the expression pattern of genes provides critical information about their temporal and spatial action and thus is a first step in understanding gene function. The most straightforward method to visualize gene expression patterns is *in situ* hybridization (ISH) on tissue sections. The sole prerequisite for carrying out ISH is having tissue preserved by methods that maintain the integrity of mRNA and having a fragment of expressed sequence. Such fragments can readily be prepared by subcloning of previously isolated cDNA clones or by reverse transcriptase polymerase chain reaction (RT-PCR) methods.[1,2]

The principle underlying ISH is to anneal a labeled DNA or RNA probe to the cellular RNA made accessible either by sectioning of the tissue of interest or by making whole mounts permeable to the probes.[3-6] ISH is very sensitive, in particular, when antisense RNA probes are used. The procedure consists of four steps: (1) the preparation of tissue (fixation, dehydration, embedding, and sectioning), (2) the preparation of the probe, (3) the application and annealing of the probe and subsequent removal of the nonhybridized probe by RNase digestion and washing, and (4) the visualization of the expression pattern (autoradiography, color reaction). Riboprobes used in ISH experiments are either labeled with ^{35}S-UTP or with a hapten (most frequently digoxigenin). In the first case, hybridization is detected by autoradiography, while in the second case antibodies are used to visualize the hapten. While sometimes offering less contrast than radiolabeled probes, digoxigenin probes reveal transcripts within the cytoplasm and thus provide cellular resolution. In the case of radiolabeled probes, the signal is detected in an overlaying autoradiographic emulsion in the form of scattered silver grains. Only if the expressing cells are large can a direct association between the cells and the silver grains be made. One advantage of digoxigenin-tagged riboprobes over radiolabeled probes is the speed by which expression data can be detected. While autoradiographs may take exposure times of 3 to 14 days, digoxigenin-tagged probes visualized by antibodies coupled to alkaline phosphatase or horseradish peroxidase can be revealed within a few hours. The protocols we chose were used successfully in our laboratories using embryos and tissues from mouse,[7,8,14] chicken,[9] and human.[7,10,11-13] In the various protocols, we usually refer to mouse or mouse embryonic tissues, both of which are readily available. Moreover, there are now many mouse models of human disease, and these animals provide a more convenient tissue source than human autopsy tissue.

Essential to all ISH analyses is the use of controls that demonstrate probe specificity. A commonly accepted control is the use of sense riboprobes. The hybridization of antisense and sense riboprobes in parallel experiments and under identical conditions is particularly important if a gene appears to be broadly expressed (Figures 5.1A,B,E, and F). Once hybridization and washing conditions are established, hybridization with the sense control may be omitted; however, when a new tissue or a different fixation procedure is used, then the use of sense probe controls is again highly recommended. It is difficult to determine which portion of a cDNA is most suitable for hybridization. If the gene of interest is a member of a gene family, we advise the use of a riboprobe complementary to the 3′ untranslated region, which is usually not conserved. However, a riboprobe designed to recognize the protein coding region gives good results in most cases, since the protocols described here use an RNase digestion step followed by a stringency wash. Even a few mismatches are sufficient for RNase to cleave the probe, leading to the removal of the cleaved probe fragments during the stringency wash. This chapter is designed to discuss each of the steps in sufficient detail in order to enable the reader to carry out ISH. An overview in Figure 5.2 presents the five steps of the procedure, the time required for each step, and the various alternatives that are available for a particular step. According to this scheme, it should be possible to generate expression data within a week's time. With some experience, the various steps can be carried out in parallel. For example, sectioning can be performed simultaneously with riboprobe synthesis. Moreover, tissues or even sections can be pre-made and stored over considerable time. In our laboratory, it is standard for an individual to carry out two series of ISH experiments per week with about 20 slides per session. During all procedures in which unhybridized tissue is handled (specimen collection, fixation, embedding, sectioning, and hybridization), it is necessary to wear gloves to avoid RNase contamination. Appendices A and B list reagents, supplies, and equipment required for ISH. We also provide names for suppliers and order numbers for items currently used in our laboratories; however, we have not systematically tested the quality of reagents from different suppliers and thus materials of equivalent grade and purity from other suppliers may be used.

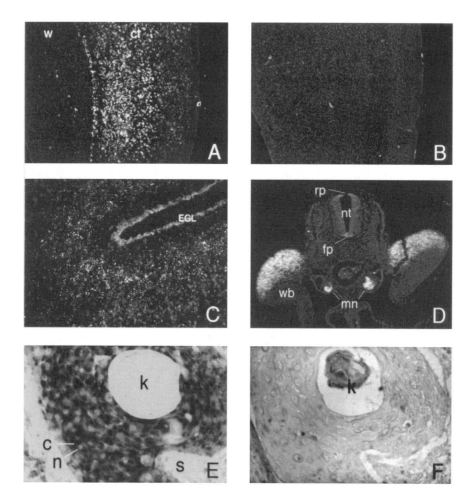

FIGURE 5.1

Examples of gene expression patterns detected by *in situ* hybridization. (**A to D**) *In situ* hybridization experiments with ^{35}S-labeled riboprobes. (**E, F**) *In situ* hybridization with digoxigenin-labeled riboprobe. (**A**) Expression of the *Lis1* gene in the cerebral wall of the mouse brain.[7] A paraffin section was hybridized with 260-nt long antisense riboprobe, and relatively broad expression of this gene (white dots) was detected in the cortex (ct). No expression is detected in the white matter (w). (**B**) Section adjacent to that shown in (A) hybridized with murine *Lis1* sense riboprobe (same length as antisense probe); no signal is detected. In (A) and (B) the tissue was visualized by staining with Hoechst 33258 fluorescent dye which binds to the DNA in the nucleus. In the black-and-white illustration, the fluorescence appears gray; however, in a fluorescent microscope equipped with appropriate filters, the nuclei are brilliantly blue. (**C**) Coronal section through the developing cerebellum of a 19-week-old human fetus. This paraffin section was hybridized with a human *Lis1* antisense riboprobe[7] 535 nt in length. Transcripts are found in a broad zone containing granule cells which are migrating from the external granular layer (EGL) to the internal granular layer. Tissue was visualized as discussed for parts (A) and (B). (**D**) Transverse paraffin section through a stage-21 chicken embryo at the level of the forelimbs. This section was hybridized with 3-kb long antisense riboprobe derived from the *lmx-1* gene. Note the strong and localized expression in the dorsal half of the wing buds (wb) and in the mesonephros (mn). There is weaker expression in the floor

II.   Tissue preparation

A.   Fixation of tissues

The general purpose of fixation is to preserve good histological detail. Because the strength of the hybridization signal is dependent upon the availability of cellular mRNA that has been preserved in the tissue throughout the procedure, fixing protocols that cross-link excessively are not suitable. Glutaraldehyde and formaldehyde are particularly suitable for ISH protocols. It is important to keep in mind that not all tissues require the same amount of fixation and that steps in the protocol are interdependent. For example, shorter fixation times or the use of gentle fixatives usually require a shorter protease digestion. A good and popular fixative for most embryonic and adult tissues is 4% paraformaldehyde in phosphate buffered saline (PBS). Fixations are usually performed at 4°C, with gentle agitation. All dissections are performed in sterile, ice-cold PBS. The tissue is rinsed in fresh PBS after dissection is complete and placed into the chosen fixative for approximately 12 hr. Convenient containers for fixation include glass scintillation vials (20-ml volume) or, if smaller tissues are being fixed, 24-well plates. While fixations of larger tissue pieces can be extended to days, care should be taken not to fix an excessive amount of tissue in a single vial. A useful guideline is that the volume of the fixative should be ≥30 times that of the volume of tissue to be fixed. Tissue destined for cryostat sections can be fixed as described above but need not be fixed when fixation following sectioning is planned (see Section II.B.2 below).

Alternative fixation agents include Bouin's or Carnoy's reagent, which have been used successfully in our laboratory for ISH of sections. Bouin's fixative is particularly useful when calcified tissues must be sectioned (older embryos), since this fixative softens hard tissues. In the case of Carnoy's and Bouin's fixative, contents in the amniotic fluid become very solid and are hard to remove after fixation. Thus, embryos should be washed in saline and extra embryonic membranes should be removed before fixation.

1.   Working with tissues from a tissue bank

Tissue from a tissue bank can also be processed for ISH. Best is tissue that was snap-frozen in liquid nitrogen. The freezing should take place immediately (≤ 5 min) after dissecting out the tissue; otherwise, mRNA is degraded and low abundance transcripts

---

plate (fp) and roof plate (rp) of the neural tube (nt). Tissue was visualized as discussed for parts (A) and (B). (**E, F**) Sections through a head and neck squamous carcinoma are shown. (**E**) Hybridization with an antisense riboprobe for RARγ (600 nt long). The dark area represents hybridization signal in the cytoplasm (c) of tumor cells; as expected, there is no signal in the nuclei (n). Stroma (s) and the keratin pearl (k) do not express RARγ. (**F**) Hybridization with sense riboprobe for RARγ as control; no signal is detected. Digoxigenin-labeled probes allow single cell resolution; compare parts (E) and (C).

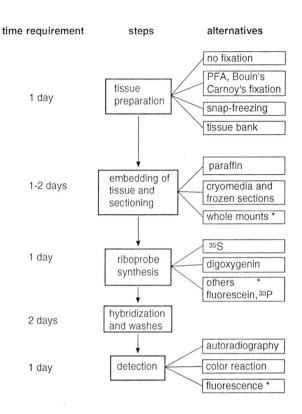

FIGURE 5.2

Diagram showing the organization of the *in situ* hybridization procedure. Each of the steps has several alternatives that are more or less suitable to attain a particular goal. Procedures marked by an asterisk are not described in this chapter.

can no longer be detected. Frozen sections that are prepared from such specimens need postfixation[15] (see Section II.B.2, below). Loda et al.[16] have used formalin-fixed and paraffin-embedded human tumors. As noted above for frozen tissues, it is critical to make sure that paraffin-embedded specimens obtained from tissue banks were rapid fixed; otherwise, RNA will be degraded.

2.   Recipes

4% paraformaldehyde/PBS: 100 ml of paraformaldehyde (PFA) fixative are prepared by heating 90 ml H_2O and 100 μl 2-*M* NaOH in a microwave oven to 50°C. The solution must be basic to aid in the dissolution of the PFA. Add 4 g of PFA (use a face mask) and stir in a hood until the PFA is dissolved. Add 10 ml 10× PBS, filter through a Whatman paper filter, and let cool to room temperature. If necessary, adjust the pH back to 7.4 by the addition of 100 μl 2-*M* HCl. If the PFA is too acidic, it will destroy DNA; if it is too basic, it will destroy RNA. Store the solution at 4°C, sealed with parafilm to prevent fixation of other proteins stored in the refrigerator. If kept refrigerated, the solution is usually good for about one week.

Carnoy's fixative: 100% EtOH (60 ml), chloroform (30 ml), acetic acid (10 ml).

Bouin's fixative: Saturated picric acid aqueous (75 ml), formalin (25 ml), acetic acid (5 ml). Suitable for larger embryos, since this fixative decalcifies the ossified tissue. To fix mouse embryos older than 17 dpc, either the skin has to be removed or the tissue of interest should be isolated by dissection from the rest of the body for better penetration of fixative. The skin of the embryo from this stage onward is impermeable to the fixative.

B. Tissue Embedding and Sectioning

1. Paraffin sections

During dehydration, water is removed from the specimen so it can be embedded in paraffin. Moreover, in a dehydrated state, specimens can be stored at –20°C for months without noticeable RNA degradation. Tissue is taken through a graded ethanol/salt series, beginning with physiologic salt concentration and ending in 100% ethanol (EtOH). The length of each wash is dependent upon the specimen size. Specimens measuring a few millimeters require 10 to 20 min per step. Specimens in the range of 5 to 10 mm require about 60 to 90 min in each step. Very large tissues should be dehydrated for 2 to 4 hr per step, with the 70% EtOH step being extended overnight. This may have to be adjusted if the tissue is less permeable. Transferring the specimen through the dehydration steps too rapidly results in specimens that cannot be sectioned. If water is not completely replaced by ethanol, or if the wax is too hot or the wax permeation period too short, air bubbles develop and appear as holes in the paraffin sections, which results in tearing of the section.

To dehydrate, subject tissue to the following steps at 4°C on a rocking platform moving at a gentle rate (20 times per min):

1. 0.9% NaCl
2. 30% EtOH in 0.9% NaCl
3. 50% EtOH in 0.9% NaCl
4. 70% EtOH in H_2O
5. 90% EtOH in H_2O
6. 100% EtOH

At this point, tissue can he stored indefinitely at –20°C.

For dehydration with Carnoy's fixative, transfer Carnoy's fixed tissues to 100% EtOH for 15 min at 4°C (twice), and store in 100% EtOH at –20°C. Transfer Bouin's fixed tissues to 70% EtOH for 15 min at 4°C. Wash in 70% EtOH and replace until the yellow picric acid is gone; otherwise, this agent crystallizes in the tissue. Transfer specimen to 95% EtOH and then to 100% EtOH for 15 min at 4°C (each twice), and store in 100% EtOH at –20°C.

If the specimen is very small (e.g., early embryos), it can be stained in eosine following step 6 above. Stained specimens can be readily oriented. Prior to embedding, immerse specimen in 1% eosine in 100% ethanol for 5 to 10 min, then rinse in 100% ethanol to remove excess dye.

Paraffin wax should be melted ahead of time, either by placing a container in a warming oven overnight, or by microwaving the wax for approximately 10 min and leaving a small amount of wax unmelted. Care should be taken not to overheat the wax, since this destroys its ability to form the proper lattice structure when the wax cools. Discard unused wax after 2 to 3 days. A small aliquot of the wax should be placed in a separate beaker, and approximately 10 min before use an equal amount of xylene is added. Discard the unused portion of the xylene-wax solution immediately in a receptacle in a hood (xylene is toxic).

Prior to embedding, process the tissues through the following solutions on a rocking plate, using the same length of time as was employed when transferring tissue from 0.9% NaCl to ethanol:

1. 100% EtOH, at room temperature.
2. EtOH/xylene, 1:1, at room temperature. Alternatively, place specimen into 100% methylbenzoate and let it sink. Determine the time required for the specimen to sink.
3. Xylene, at room temperature; leave specimen in xylene for the time that was required for the specimen to sink in the 100% methylbenzoate (step 2).
4. Xylene-wax, at 58°C (does not need to be rocked).
5. Wax, 3 times at 58°C.

After the final wax wash, prepare a suitable container for the tissue, such as a small weigh boat or a commercially available plastic mold. Gently swirl the vial to suspend the tissue in the melted wax, and quickly pour into the container; avoid creating bubbles in the process. The tissue can be oriented using forceps that have been warmed over an alcohol flame. If the tissue fails to pour from the vial, add more wax and try again. Bubbles in the molten wax can be removed by heating the forceps and gently moving it through the wax. Bubbles that are in close proximity to the tissue should be removed, as they will create voids in the wax that will make sectioning more difficult. The wax blocks should be allowed to cool at room temperature without disruption, and can be stored at room temperature indefinitely. Do not refrigerate or freeze wax blocks.

Tissue fixed with Bouin's fixative and stored in 100% EtOH is embedded as follows. Transfer specimen to methylbenzoate with three exchanges at room temperature; after the last change, wait until the specimen sinks to the bottom of the vial. Transfer to benzene (twice, room temperature), benzene wax (1:1, 58°C), wax (twice, 58°C), and pour into mold. Methylbenzoate and benzene are used instead of xylene since the latter hardens the tissue and makes it too brittle for sectioning.

Mounting the tissue onto a wooden block (or an equivalent support) is easily performed if there is sufficient wax around all sides of the embedded tissue. Determine the plane of section that is desired, then spread a small amount of melted wax onto the block and heat up a metal spatula until it is very hot. Place the warmed spatula against the wax on the block and simultaneously lower the tissue-containing wax block towards the top surface of the heated spatula. Quickly remove the spatula and firmly place the now-melted surface of the wax block into the heated wax on the wooden block. Reheat the spatula and melt the four sides of the wax block slightly, allowing the dripping wax to fill in any space between the wooden block and the tissue-containing wax block; allow sample to cool at room temperature.

In order to produce a ribbon of tissue sections that is straight and wrinkle-free, the wax block must have 90° angles. If the bottom edge is uneven or either side is longer than the other, the ribbon will twist, making it difficult to position the sections onto a slide. Gradually shave down the wax block around the tissue with a clean razor blade, preserving enough wax at the periphery so that the tissue occupies approximately one half of the total wax surface. Take care to remove only small amounts of wax at a time so that the block does not crack.

Sections are best cut in a place with few air currents and constant temperature. Position a water bath (set between 45 and 50°C, use autoclaved water) and a slide warmer set at 38°C in close proximity. The position and angle of the microtome blade, the speed at which to cut, and the optimal thickness of the section are determined empirically. To begin, use a blade angle of 10°, 7-μm thickness, and a slow cutting speed. If the sections tend to wrinkle, it may be due to compression as a result of cutting too fast; thus, slow down when passing through the tissue. In addition, one can also reduce the angle of the knife. If the sections fail to form a ribbon and come off as individual sections, it is frequently due to the angle of the blade; adjust blade to a larger angle. If you detect a scratch or a split in the wax section, first check that there are no obvious nicks in the blade. If so, reposition the microtome blade so that a new zone is being used; use extreme caution when doing this so as not to cut yourself. Dust particles in the block underneath the specimen also can cause failure to form a ribbon. Shave a small amount of wax off the bottom edge of the wax block (the edge where the knife first contacts the block).

Once you have succeeded in making ribbons of sections, lay them onto a clean piece of paper. It is particularly helpful to use black paper since you can see the tissue in the wax with more contrast. After you have accumulated a number of ribbons, you can begin to mount them. Cut the ribbons into sections that are shorter than the slide you are using; the ribbon will expand approximately 10% as it warms up.

Place the ribbon gently into the water bath, taking care to place the side of the ribbon that faced the blade in contact with the water (shiny side of ribbon). Watch the ribbon closely; as the wax melts and expands, small wrinkles should disappear. Do not allow the water to get too warm, as the wax will begin to melt, and in doing so will tear the tissue. To mount the tissue, position a slide underneath the ribbons and, using forceps to guide the sections, slowly lift the slide out of the water. Care should be taken not to touch the tissue since the wax will melt slightly and become adherent to the forceps. Allow the slides to dry on the slide warmer for a minimum of 2 hr, then store them in a cool (4 to 20°C), dry place. Include a desiccant in the box, then seal the box. Avoiding water condensation is critical, since in the presence of water RNase can start cleaving the cellular RNA.

Hints and problems

1. Minimizing the length of the xylene step is advisable, since extended exposures cause tissue to become brittle (see also alternative method with methylbenzoate mentioned above).

2. *Overheated wax shrinks after it has been poured, creating a depression in the wax block.* Do not attempt to re-pour wax into this depression, as it will not adhere to the original wax, causing separation when the wax block is mounted for sectioning. Instead, the entire wax block is melted by placing it back into the oven for the minimum time necessary to completely melt the wax. Re-pour and let cool to room temperature.

3. *Poor tissue quality on sections.* Brittle tissue which cracks upon sectioning usually indicates excessive time in xylene or too high temperature of the wax. If air bubbles are associated with the tissue, it is an indicator that the tissue is being moved too rapidly through the ethanol-xylene-wax solutions. Using old wax can also result in poor tissue embedding.

4. *Wrinkled sections, sections coming off of the slides.* Wrinkled sections frequently result when the wax block is cut too quickly or static electricity prevents the ribbon from moving away from the blade. Either slowing down the speed with which the blade passes through the knife or wetting a Kimwipe and touching it to the wax block to decrease static electricity usually helps. In very dry rooms, air humidification may be required to reduce static electricity. Sections can come off the slides during the protocol when they were wrinkled during sectioning. In addition, inadequate postfixation with PFA may be the cause; make sure to use fresh PFA.

Recipes

There are a number of protocols for treating slides to aid in adherence of the sections. We present two methods, one of which cleans and etches the slides slightly using TESPA (3-aminopropyltrioxyethyl silane); these slides are stable indefinitely. The other method uses poly-L-lysine; although the tissue adheres very well, poly-L-lysine is labile and the slides are more difficult to prepare.

TESPA: Place slides in a stainless steel rack and dip in 10% HCl/70% EtOH for 2 min, followed by distilled water and then 95% EtOH, 2 min each. Dry the slides either in a baking oven at 150°C for 15 to 20 min or in Speed-Vac, then cool to room temperature. Dip slides for 10 to 15 sec in 2% TESPA in acetone, followed by two quick washes in acetone and one wash in water. Dry at 42°C and place in a vacuum or desiccated box.

Poly-L-lysine: Place slides in a stainless steel rack and wash in a 1% solution of Linbro ("7X") for 15 min. Rinse the slides extensively in deionized water (5 times, 2 to 3 min each), then bake them at 150°C for 1 hr. Cool to room temperature before immersing the slides in a freshly prepared solution of 100 μg/ml poly-L-lysine in autoclaved deionized water for 1 hr at room temperature. Allow the rack to drain and air dry for a few hours. Store in a vacuum or desiccated box for up to 2 weeks.

Eosine (1% stock of alcoholic eosine): 1 g of Eosin Y is dissolved in 100% EtOH.

2.   Cryosections

Similar to paraffin sections, the preparation of frozen sections consists of several steps: fixation, freezing of tissue, embedding and mounting into a cryomedium, sectioning in the cryostat, and placing on a slide. To fix tissue, place it into 4% paraformaldehyde prior to freezing as described in Section II.A. Wash in cold PBS, place tissue into a small, self-made aluminum foil cup, add embedding medium (Histo Prep or OCT), and snap-freeze in liquid nitrogen or isopentane chilled on dry ice. Unfixed tissue is embedded the same way, but, once sectioned, unfixed tissue requires a postfixation for ISH to be successful (see below). Frozen specimens can be stored at −75°C. Tissue morphology is improved if the sample is permeated with sucrose. 4% PFA fixed tissue is briefly washed in PBS and transferred to 0.5 M sucrose (in DEPC-treated PBS; do not autoclave sucrose solution). Gently shake on a platform for 1 to 3 hr at 4°C, until specimen has settled. Place specimen onto a tracing paper, cover with a bit of cryomedium, and freeze on top of a slab of dry ice.

Snap-frozen but not yet embedded tissue is embedded in cryomedium as follows. Cool 100 ml isopentane (2-methylbutane) for 20 min on dry ice (do not cool longer). Place a small amount of cryomedium onto the metal chuck of the cryostat (the chuck has to be at room temperature; using a cold chuck causes the cryomedium to pop off during sectioning), using enough medium to build a platform approximately 0.5 to 1 cm thick. Using forceps, put the metal chuck into the isopentane for a few seconds to partly freeze the embedding medium The surface of the medium should remain unfrozen. Quickly place the frozen tissue block in the desired orientation into the medium and add more medium to cover the specimen. Immediately immerse the chuck and the mounted tissue in the isopentane for 3 min. If the tissue is already embedded in cryomedium, the block is mounted onto the metal chuck in the same manner, except that no additional embedding medium has to be used to cover the block. Mount the metal chuck immediately on the cryostat holder and let the sample equilibrate to the cryostat temperature for at least 15 min.

Sectioning involves the following steps. Trim the block with a cold razor blade to generate a cutting face that is only slightly larger than the tissue. Adjust the angle of the knife to 5 to 15°. The angle has to be determined empirically depending on the thickness of section (usually 10 to 15 µm) and the embedding method. With the anti-roll plate clear of the knife, cut several sections at the desired thickness and observe their quality and uniformity. When satisfactory sections have been cut, clean the knife surface with a cold bristle brush. Flip the anti-roll guide plate so it rests on the knife surface. The tissue sections should now pass smoothly between the knife and the anti-roll guide plate; generally, a slow operating stroke provides the best results. Swing the anti-roll guide plate clear of the knife without haste to minimize disturbance to the flattened section. Take a TESPA-coated slide (room temperature) and bring it up to and parallel to the section on the knife. The section should now transfer automatically onto the slide. Take slide out of the cryostat chamber and let the slide dry (≈1 min) to remove condensation. Slides with sections can be stored at −70°C. For further processing of tissue fixed prior to sectioning, proceed to step 6 in Section III.B.1. To fix tissue after sectioning, immerse slides in 4% PFA for 30 min. Subsequently, wash the slide twice in PBS to remove traces of embedding medium. This material would bind to the riboprobe, resulting in a high background. In the case of postsectioning fixed tissue, proteinase K treatment is not necessary; thus, proceed to step 11 in Section III.B.1.

Hints and problems

1. *Tissue block becomes detached from object holder.*
   a. Tissue is not frozen enough when mounted.
   b. Object holder had dirt or grease on platform.
   c. Bottom surface of tissue block is not flat enough.
   d. Slices are being trimmed too thick.
   e. Tissue is below knife when advancing feed.
   f. Knife clearance angle is insufficient

2. *Thickness of successive sections is variable.*
   a. Irregular advance of microtome feed could be due to ice on feed screw or ratchet wheel.
   b. Object holder or knife is not clamped securely.
   c. Microtome rocker knife edges are subject to excessive icing.
   d. Clearance angle of knife is insufficient
   e. Tissue is being compressed due to tilt angle of knife being too great.
   f. Knife edge is blunt
   g. Tissue expands due to warming up of block.
   h. Block is being sectioned when it is colder than the microtome chamber, such as happens when cutting immediately after freezing.
   i. Interval between section cutting is too long, allowing small temperature change to reduce dimensions of block.
   j. Tissue is insecurely attached to object holder.

3. *Sections crumble or do not form.*
   a. Knife is blunt.
   b. Angle of knife is wrong.
   c. Tissue, knife chamber, or anti-roll plate is too warm.
   d. Tissue is dehydrated.

4. *Sections compress.*
   a. Guide plate or knife is covered with ice or frozen pieces of tissue.
   b. Angle of knife is wrong.
   c. Knife is blunt.
   d. Knife or guide plate is too warm or dirty.

5. *Horizontal cracks are found in section.*
   a. Tissue block is too cold.
   b. Tissue block was frozen too slowly.

6. *Vertical scores or lines appear in sections.*
   a. Ice or frozen tissue is on edge or face of guide plate.
   b. Top edge of guide plate is damaged.
   c. Knife edge has been damaged or has nicks.
   d. Knife edge or face is covered with pieces of ice or frozen tissue.

7. *Tissue shatters.*
   a. Tissue block is too cold.
   b. Tissue block is dehydrated.

8. *Section sticks to guide plate.*
   a. Guide plate is not cold enough.
   b. Guide plate is dirty or greasy.
   c. Guide plate angle to knife is incorrect.

9. *Flattened section curls up upon removal of guide plate.*
   a. Angle of guide plate to knife is too great.
   b. Guide plate removal from knife was too rapid, causing air current to disturb section.

III.   Hybridization with ^{35}S Riboprobes

A.   Preparation and Synthesis of ^{35}S Riboprobes

Single-stranded RNA probes are prepared using plasmid vectors containing a polylinker bordered by promoters of T3 and T7 bacteriophage RNA polymerases. Probe synthesis reactions are carried out using a linearized plasmid as a template. The plasmid is linearized by digesting with a restriction enzyme that cuts at a single site downstream of the insert. The site should be oriented so that the synthesized RNA probe will be terminated as a result of polymerase runoff. In order to synthesize a probe that hybridizes to mRNA, the antisense strand of the cloned insert must be used as the template. Following restriction enzyme digestion, the template is treated with proteinase K, then extracted in phenol/chloroform and precipitated in ethanol. This proteinase K treatment removes ribonucleases from the template reaction. The linearized template is then resuspended in diethyl pyrocarbonate (DEPC)-treated water or TE buffer (10 mM Tris Cl; 1 mM EDTA, pH 7.4).

1.   Template preparation

cDNAs should be digested to yield both sense and antisense orientations of the template. The sense probe serves as an important control for nonspecific hybridization. At least 10 μg of DNA should be digested initially. Determine that the restriction digestion has gone to completion by running a small aliquot of the reaction out on an agarose gel; the presence of the uncleaved template will generate RNA transcripts that contain polylinker and vector sequence and may be a source of background signal. Add 50 µg/ml of proteinase K to the restriction buffer and incubate for 30 min at 37°C, followed by two phenol/chloroform (1:1 v/v) extractions and ethanol precipitation. Resuspend the digested, proteinase K-treated cDNA in TE made with DEPC-treated water.

2.   Transcription reaction

A typical transcription reaction is described which will generate high-specific activity probes (transcription kits are commercially available). Add the solutions in the following order, keeping the reagents on ice and using caution to avoid contamination with RNases (gloves, autoclaved tubes and tips, etc.).

> 6 µl of 5× transcription buffer (5× is 200 mM Tris-HCl, pH 8.0; 40 mM MgCl$_2$;
>   250 mM NaCl; 10 mM spermidine)
> 1 µl of 1 µg/µl of restricted, proteinase K-treated, linearized DNA template
> Add sufficient DEPC-treated water to make a final volume of 30 µl
> 1 µl of 10-mM rATP
> 1 µl of 10-mM rCTP
> 1 µl of 10-mM rGTP
> 1 µl of RNAsin (40 U/µl)
> 1 µl of 0.75-M DTT (use DTT that has been thawed once)
> 5 to 10 µl of 1000-Ci/mmol, 10-µCi/µl ^{35}S-UTP

Mix well, spin, and then add 10 U of T3 or T7 RNA polymerase. Incubate at 37°C for 2 to 3 hr.

To remove the template DNA, use DNase I that is RNase-free, and add the following to the 30-µl transcription reaction: 19 µl of DEPC-treated water, 1.7 µl of 0.3-M MgCl$_2$, 2 U of DNase I. Incubate at 37°C for 15 min.

The probe is precipitated by adding the following to the transcription reaction: 100 µl of DEPC-treated water, 100 µl of 1-mg/ml yeast tRNA (previously purified by phenol extraction), 250 µl of 4-M ammonium acetate, and 1 ml of EtOH. Vortex well, precipitate 5 min on ice, spin 10 min, and remove supernatant. This precipitation is repeated once. Aspirate off the remaining supernatant using a pulled-out Pasteur pipet. A Speed Vac can also be used for this purpose; however, use caution, since RNA pellets may be difficult to resuspend if overdried. Resuspend pellet in hybridization buffer (see Section III.B). Alternatively, unincorporated nucleotides are removed by passing the transcription reaction over a column containing RNase-free Sephadex G-50.

If hydrolysis is necessary to reduce the size of the transcripts (longer than 1 to 1.5 kb, with the optimal probe size being 0.5 to 0.8 kb), dissolve the probe in water and add an equal volume of 80-mM NaHCO$_3$, 120-mM Na$_2$CO$_3$. Incubate at 60°C for t minutes, where t = (L_o – L_d)/0.11 L_o L, where L_o is the original probe length in kilobases and L_d is the desired probe length.

3.   Determination of incorporation

Typically, greater than 50% of the input ^{35}S-UTP is incorporated into the RNA transcription reaction. The use of labeled UTP as the single source of this nucleotide limits the mass yield of RNA probe. The following calculation is used to determine the amount of RNA generated in the transcription reaction:

$$\frac{\text{Total number of }\mu\text{Ci added} \times 13.2 \times \text{\% incorporation}}{\text{Number of }^{35}\text{S-dNTP used} \times \text{specific activity of radioisotope}} \quad = \quad \text{ng of RNA synthesized}$$

Percent incorporation is determined by total counts divided by incorporated counts. Count an aliquot of riboprobe and an aliquot of the reaction mix taken prior to precipitation in 2 ml water-miscible scintillation fluid in the ^{14}C channel. The approximate specific activity of freshly prepared RNA is expected to be 10^9 cpm/μg of RNA. If necessary, store the probe at –80°C for up to a week (fresh probes are better since older probes display strand breaks).

B.   Hybridization Steps

1.   Dewaxing and postfixation of sections

All solutions used for prehybridization should be RNase-free. This is achieved by autoclaving for 45 min. It is convenient to prepare 10× stock solutions and make dilutions as required. Most of the prehybridization solutions can usually be reused if they are kept RNase-free. Paraformaldehyde (PFA) solutions can be reused twice, but should be discarded thereafter. PFA is stored at 4°C. Proteinase K and acetylation solutions must be made fresh each time. We use a Tissue-Tek II staining station from Miles for the steps described here (see Appendix B). It is advisable to cover all solutions containing ethanol with plastic wrap to prevent evaporation. Place slides (in a rack) in the following solutions in the order listed:

1. Histoclear or another xylene substitute to remove wax, 10 min, twice
2. Ethanol series, starting at 100% and progressing to 95, 70, 50, and 30%; agitate the slides in each solution until equilibrated (approximately 1 min each)
3. 0.9% NaCl, 5 min
4. 1× PBS, 5 min
5. 4% PFA, 20 min (this solution can be reused below)
6. 1× PBS, 5 min
7. Subject slides to proteinase K treatment: 20 μg/ml proteinase K in 50 mM Tris-HCl, pH 7.6; 5 mM EDTA, 5 min (after this step, the tissue is particularly susceptible to RNases; use extreme caution)
8. 0.2 M HCl, 5 min (can be left out)
9. 1× PBS, 5 min
10. 4% PFA, 20 min
11. Acetylation: The purpose of this step is to acetylate amino groups in tissues to reduce electrostatic binding of probe. Acetylation also blocks binding of probe to poly-L-lysine-treated slides. These steps must be performed in a hood or another well-ventilated

place. Suspend the slide rack above a rapidly rotating stir-bar in a solution of 0.1 *M* triethanolamine-HCl (TEA, pH 8.0). Add 600 µl of acetic anhydride; wait 5 min, then add 600 µl more of acetic anhydride. Incubate for a total of 10 min. Care should be taken not to introduce water into the acetic anhydride stock; acetic anhydride would rapidly hydrolyze to acetic acid in the presence of water. Thus, if the agent smells like acetic acid, it should be discarded.

12. 1× PBS, 5 min

13. 0.9% NaCl, 5 min

14. 30% and 50% EtOH, 1 min each

15. 70% EtOH, 5 min (this step is longer to ensure removal of all salt deposits)

16. 80, 95, 100, and 100% EtOH, 1 min each

Air dry slides in a dust-free, RNase-free place, then store in a sealed container at room temperature with desiccant for up to a few days. Theoretically, dewaxed slides can be stored indefinitely; however, they are much more susceptible to RNase once the tissue has been treated with proteinase K and are also prone to oxidation.

2.   Prehybridization

Sections may be incubated with hybridization buffer not containing the probe in order to reduce a high background; this may help especially if the mRNA of interest is rare. For 25 × 75-mm slides with 24 × 50-mm coverslips, 100 µl of hybridization mix per slide is sufficient. Be sure to prepare extra mix, because the viscosity of the hybridization mix leads to large pipetting errors. To the mix, add 1/100th volume of 25-m*M* alpha-S-thio ATP (this is crucial for the reduction of background). The mix is then heated in 90 to 100°C water for 2 min and cooled to room temperature. Carefully place the hybridization mix onto the dry slide, then lower a coverslip onto the slide with forceps. This is easiest if you drop one end of the coverslip onto the slide, then slowly lower the rest. Try to avoid bubbles; if they do form, make sure that hybridization mix completely covers every section. Incubate slides in horizontal position for 1 hr at 50°C in a chamber whose bottom is filled with several layers of Whatman paper soaked in 50% formamide/2× SSC. We have used chambers consisting of transparent plastic obtained in hardware or department stores. We place the slides on a grid rack built from plastic pipets.

Prehybridization has disadvantages. First, it necessitates removing the coverslip covering the section which may damage the tissue. Second, the exact probe concentration cannot be determined for the subsequent hybridization, since hybridization buffer remains on the slide and will dilute the probe. The alternative, a hybridization performed at a higher temperature, can also decrease background and has the advantage of being quicker and not requiring the removal of the coverslip.

Hints and problems

In the case of low transcription from the template, first check that the RNA pellet is completely resuspended by vortexing and heating slightly (50°C). Second, check that the appropriate polymerase was used for a given template. Also, check whether the probe is intact by running a small aliquot out on a gel (using care to prevent RNA degradation by RNases). Lastly, extensive amounts of secondary structure which permits intramolecular annealing in the template DNA may prevent a good yield of high specific activity probe. In this case, a better probe may be obtained by using a shorter template or by using another region of the template.

3.   Hybridization

An optimal signal-to-noise ratio is achieved when probe concentrations are just sufficient to saturate the complementary cellular mRNAs; however, this concentration is usually unknown, and the optimal probe concentration is therefore determined empirically. Initially, the amount of probe used in a hybridization reaction should be in 10-fold or greater excess of the transcripts of interest. A good starting place is to use a high specific activity probe (10^9 cpm/µg riboprobe) at a concentration of 2 to 6×10^6 cpm/slide (2 to 6 ng). We have found, however, that using half or even less (1/10th) of that amount can yield even nicer *in situ* when the expression of the gene of interest is confined to a small area. The correlation is not intuitive; it seems that the lower the expression level of the mRNA species, the lower the optimum concentration of the probe. It is probably a good idea to start off with the 1 ng per slide and then subsequently titrate to find the optimal concentration. The rate of hybridization depends on the concentration of probe. We found that overnight hybridizations are sufficient, although times as short as 5 hr have been successfully used.

If the sections were prehybridized, remove the coverslip by tilting the slide and simply letting it slide off. A 26 × 75-mm slide can be covered adequately with 100 µl of hybridization solution (50 µl are sufficient for prehybridized slides); calculate the appropriate amount of probe to add to the hybridization buffer, and place into an Eppendorf tube. Place the tube into a 70°C water bath for 3 min. It is not necessary to denature RNA probes prior to hybridization; however, heating the hybridization buffer prior to pipetting is helpful due to the viscous nature of the solution. Apply the mix with a Pipetman and gently lower a coverslip onto the slide using forceps by decreasing the angle between the coverslip and slide so that bubbles are forced out. The coverslips must be very clean to minimize bubbles. Do not press down on the coverslip to remove bubble, as this will result in a thinner layer of hybridization solution. Place the slides into the humidified, sealed chamber (see Section III.B.2, above). Incubation temperatures are determined empirically; a good starting point is 40 to 50°C, which is approximately 25°C below the melting temperature (T_m) of hybrids formed *in situ* with a 150-nucleotide-long probe and a GC content of 50%.

4.   Posthybridization washes

Washes are used to gently remove the coverslip, then to hydrolyze the nonspecifically bound probe which has adhered to the slide and the tissue. The optimal temperatures of the washes must be determined empirically; however, a temperature of 50°C will provide the same stringency as the hybridization conditions. Higher temperatures, up to 65°C, can be used to improve background without significant loss of signal. Proceed as follows:

1. Coverslip removal: gently place the slides into a slide holder, using extreme caution not to dislodge the coverslips. Do not attempt to remove coverslips manually since the shear force is likely to damage the tissue. Place the slide rack into a solution of 5× SSC at 50°C; add 40 mM beta-mercaptoethanol (in a hood!). Gently agitate the slides every few minutes; after 15 min the coverslips should have loosened sufficiently so that upon raising a slide out of the holder, the coverslip is left behind in the solution. Discard this solution in the radioactive waste.

2. Treat slides in 50% formamide, 2× SSC, 40 mM beta-mercaptoethanol at 55 to 65°C for 30 min.

3. Wash slides once or twice in 0.5 M NaCl, 10 mM Tris-HCl (pH 8.0), 1 mM EDTA, 40 mM beta-mercaptoethanol for 30 min at 37°C.

4. The nonspecifically bound probe is hydrolyzed using 10 to 20 µg/ml RNase A in 0.5 M NaCl, 10 mM Tris-HCl (pH 8.0), 1 mM EDTA for 30 min at 37°C. RNase A is stored as a stock solution of 10 mg/ml which can be refrozen. If background is a problem, the addition of RNase Tl (at 1 U/ml) is recommended.

5. Slides are re-equilibrated in 0.5 M NaCl, 10 mM Tris-HCl (pH 8.0), 1 mM EDTA without RNase A for 15 min at room temperature, then subjected again to a high stringency wash (step 2 above) of 50% formamide/2× SSC, 40 mM beta-mercaptoethanol.

6. Wash once in 2× SSC for 15 min, then 0.1× SSC for 15 min, each at room temperature.

7. Dehydrate through an ethanol series: 30, 50, 70% EtOH/0.3 M NH$_4$ acetate, then in 95% and twice in 100% ethanol. Allow slides to dry. Step 2 (first stringency wash) and step 3 can be omitted if the probe behaves well (not "sticky").

Recipe

Hybridization buffer: 50% deionized formamide, 0.3 M NaCl, 20 mM Tris-HCl (pH 8), 5 mM EDTA, 10% dextran sulfate (heat gently and rock to dissolve), 0.02% Ficoll, 0.02% bovine serum albumin (RNAse-free), 0.02% polyvinylpyrrolidone, 0.5 mg/ml yeast RNA (RNase-free), 0.1 M DTT. Make 1-ml aliquots and store at –80°C until use.

C.   Autoradiography and Visualization

To determine whether the stringency washes were of a sufficiently high temperature and to determine the length of time that slides should be exposed to emulsion, the slides are first exposed to a high-performance autoradiographic film for 1 to 3 days. Cronex film is suitable for this purpose. Tape the top edges of the slides to an immobile surface (a used piece of film is ideal) and use care not to allow the slides to come in contact with dust or lint present in the film cassette. Ideally, antisense probe hybridization should reveal distinct hybridization in the sections, possibly only in specific regions such as the nervous system, the heart, or some other subset of tissues. The sense control should be essentially free of signal except for a weak uniform haze on top of the section. If this is not the case, the temperature of the stringency wash should be increased. Some genes are broadly expressed, as would be revealed with the antisense probe. If this is suspected, it is extremely critical that the sense control be essentially blank. Sometimes there is intense signal around the margins of sections or even on the slide. This can be due to the slides drying out during hybridization; other causes are discussed in Section III.C.3.

If it is determined that a higher wash temperature is required to remove background, the slides can be rehydrated by placing them in 2× SSC for 15 min and then into a solution of 50% formamide/2× SSC/40 mM beta-mercaptoethanol. Rewashing can often remove nonspecifically bound probe; however, it rarely works as well as having washed initially at a higher temperature. After rewashing, dehydrate the slides as described under Section III.B.4, step 5.

1.   Application of emulsion

Slides of suitable quality are then dipped in photographic emulsion. Emulsion is prepared by adding 118 ml Kodak NTB-2 emulsion to 200 ml deionized water. Read directions carefully when using this emulsion; most darkroom safelights are not suitable and will expose the emulsion; dark red (number 2) safelight is suitable. The emulsion and water must be warmed at 40 to 42°C for 30 to 35 min before mixing, using extreme caution to avoid exposure to any light source. It is convenient to aliquot the emulsion into glass scintillation vials, approximately 10 ml each. Wrap the scintillation vials in two layers of aluminum foil, then place inside a light-tight box. This box should be stored at 4°C and kept away from all forms of beta radiation.

Melt an aliquot of the emulsion at 42°C for 5 to 10 min (keep this time to a minimum since the emulsion tends to break down with heat and time). Do not agitate. Once melted, gently pour into a suitable container such as a slide mailer (e.g., Young Laboratories), using care not to create bubbles. The slides are then grasped at the frosted end and gently lowered into the emulsion for 3 to 4 sec. Wipe the back of the slide to remove the emulsion and immediately place slide horizontal. The slides must be kept horizontal until the emulsion is completely hardened. This is dependent upon the airflow in the chamber in which the slides are drying. If a simple box is used, the edges usually require taping to exclude all light. A more convenient option is either to obtain a desiccant box which is airtight and place silica gel desiccant in it or to have a box manufactured. If a box is manufactured, it is convenient to have a fan installed; however, note that emulsion cannot be dried under vacuum nor can it be heated, as it tends to crack. Drying time is usually 5 to 7 hr at room temperature.

Once dry, the slides can be stored in a sealed black box containing silica gel desiccant. The box should be wrapped in aluminum foil, shielded from beta radiation, and stored at 4°C. Use the exposure time of the Cronex film as a guide to determining the length of the emulsion exposure. Usually 3 to 4 times the length of the Cronex exposure is sufficient. Batches of emulsion should be tested prior to dipping tissue slides. Typically, the silver grain density should be so low as not to create a haze. Moreover, there should be no local clusters of silver grains.

2.   Developing

Remove the slides from 4°C and let them equilibrate to room temperature. Using only the appropriate dark red (number 2) safelight, place the slides into a rack and develop as follows:

1.   Use Kodak D-19 developer for 2 min (do not warm past room temperature).

2.   Wash in water for 10 to 15 sec.

3.   Place in fixer for 5 min.

4.   Final rinse in water for 10 to 15 min.

After developing, the slides can be counterstained with a solution of 2 μg/ml Hoechst 33258 in water (stock solution of 10 mg/ml in DMSO; make 40-μl aliquots and store at –80°C). Dip the slides into the solution for 2 min, followed by 2 min in water. Air-dry the slides in a dark place, since the Hoechst stain is light sensitive. Once dry, overlay the slides with 50 μl of a solution of 5 g Canada balsam in 10 ml methyl salicylate, then place a coverslip and wick off the excess mounting medium from the edges. Keep slides flat for 1 week so that the mounting medium dries out. Other mounting media may be used which dry faster; however, some of those exhibit autofluorescence and cannot be used in combination with Hoechst staining. An alternative staining method not involving fluorescence detection involves staining with toluidine blue.[3] If a coverslip has been mounted onto the slide with Canada balsam, the coverglass can be removed by immersing the slide in methyl salicylate for 1 hr with subsequent lifting of the coverslip from the slide. After rehydration of the sections through a series of 2-min washes from 100 to 70 to 30% ethanol and water, the slide can then be subjected to hematoxilin staining.

3.   Viewing and photography

Hoechst stain is a nuclear dye which has the advantage of allowing a determination of cell density as well as visualizing tissue morphology. Using the DAPI channel on an epifluorescence microscope, at the same time as one views the silver grains by transmitted light, allows one to determine the tissue which is expressing the gene of interest. Hoechst dye eventually bleaches under fluorescent light; therefore, care should be taken to minimize exposure. To clean the slides for photography, remove residual emulsion with glacial acetic acid on a Q-tip. Remove acid with ethanol. One may also use a new razor blade and scratch off the emulsion on the back of the slide; do not scratch the glass itself since this will cause background in the darkfield. Typically, we obtain images of our *in situ* hybridizations specimens by conventional photography. In this case, we take an image of the tissue revealed by blue Hoechst fluorescence and superimpose a darkfield image. To enhance the contrast, we insert a red filter into the light path so that the silver grains become red instead of white. More convenient than classical photographs are images captured with a videocamera linked to a computer. These can be "black-and-white" images which are then pseudocolored using programs such as Adobe Photoshop. Such images can readily be stored on CD-ROMs, assembled into figures, and printed with a dye-sublimation printer. Electronically stored images can also be transmitted to other sites using computer networks.

Hints and problems

1. *High background*: This can result from a variety of problems. Too low of a wash temperature is a frequent problem. Try increasing the temperature 5°C in both the hybridization and wash steps. A major source of background signal can result when using antisense RNA probes that are contaminated with a small amount of sense-strand RNA. A common source of such sense-strand synthesis is the initiation of the RNA polymerase from the ends of the template DNA. Such aberrant initiation is more frequent when the ends of the template have a 3′ overhang. If even a small amount of the probe is double stranded, it will be resistant to degradation by RNase A. Sense RNA can be isolated away from antisense RNA by gel electrophoresis. When using ^{35}S-labeled probes, either DTT or beta-mercaptoethanol is required to prevent oxidation of sulfhydryl groups. Both agents are heat labile; therefore, it is advisable to add these compounds just before adding the slides to a heated solution, using them once and discarding the unused portion.

2. *High background in emulsion:* Usually this results from exposure of the emulsion to light or beta radiation. Check that the appropriate safelights are being used and the box in which the slides are being dried is light-tight. Be aware of watches that glow in the dark or beepers with lights. The distance between the safelight and the emulsion should be a minimum of 4 ft.

3. *Emulsion removal:* At times it may be necessary to remove the emulsion layer from the slide — for example, if the unprocessed emulsion layer has been unintentionally exposed to light or the processed emulsion layer has high background and the specimens are of sufficient value to warrant recovery. Although it is difficult to remove the emulsion without some damage to the underlying tissue because of the use of alkaline solutions, the following protocols will allow removal of most of the emulsion. If the emulsion has not been hardened by photographic developing, it is reasonably soft and can be removed by placing the slides in a 60°C water bath for a few minutes to soften the emulsion. Rinse the emulsion off by gentle agitation in the water bath or under running tap water. Place the slides in Kodak fixer for 3 min to remove the faint bluish stain, followed by washing for 5 min in running water to remove all traces of the fixer. After drying completely, the slide can be redipped in emulsion. If the emulsion layer has been hardened by photographic processing, it is much more difficult to remove it. First, digest the slide for 2 to 5 min in a 1% solution of NaOH or KOH at room temperature. Any remaining silver grains in contact with the tissue can be removed by placing the slides in a silver reducer, such as 7.5% potassium ferricyanide. This is followed by a water rinse, then 3 min in Kodak fixer and 5 min of water washing. The slides can then be redipped.

IV.   Hybridization with Digoxigenin-Tagged Riboprobes

Here, we describe the use of digoxigenin-tagged riboprobes on paraffin sections. Tissue collection, fixation, and sectioning are described under Sections II.A and II.B.1.

A.   Preparation and Synthesis of Digoxigenin-Tagged Riboprobes

1.   Template and transcription reaction

It is better to have long inserts ($\geq$0.6 kb) to increase the sensitivity of detection of transcripts which are expressed broadly or expressed at low levels. The procedure described here will work well with riboprobes ranging from 0.3 to 1.2 kb.

Single-stranded riboprobe is prepared by the transcription of linearized plasmid template using digoxigenin-tagged rUTP in the substrate mix. We include 0.2 µCi of α^{32}P-rUTP as a tracer to determine the yield of the reaction.

> 4 µl of 5× transcription buffer (5× is 200 mM Tris-HCl, pH 8.0; 40 mM MgCl$_2$; 250 mM NaCl; 10 mM spermidine)
> 1.5 µl of 1 µg/µl of restricted, proteinase K-treated, linearized DNA template
> Sufficient DEPC-treated water to make a final volume of 20 µl
> 2 µl of 10-mM rATP
> 2 µl of 10-mM rCTP
> 2 µl of 10-mM rGTP
> 1.3 µl of 10-mM rUTP (can be replaced by dig-11-rUTP)
> 0.7 µl of 10-mM dig-11-rUTP
> 0.5 µl of RNasin (40 U/µl)
> 2 µl of 0.1-M DTT (use DTT that has been thawed once)
> 1 µl diluted α^{32}P-rUTP (0.2 µCi per µl)

Mix well, spin, and then add 40 U of T3 or T7 polymerase. Incubate at 37°C for 2.5 to 3 hr. Digest template by adding 2 µl of DNase I and incubate for 15 min at 37°C. Take 1 µl aliquot of synthesized probe to measure yield of incorporation (probe can be stored on ice or at −20°C in the meantime).

2.   Determination of incorporation

Spot two 0.5-µl aliquots of reaction on two Whatman GF/C glass-fiber filters and let filters air dry. Transfer one filter into 200 to 300 ml ice cold 5% TCA containing 20 mM sodium pyrophosphate. Swirl the filters in this solution for 2 min. Wash two more times. Transfer the filter to 70% EtOH, wash briefly, then let it air dry. This filter and the nonwashed filter are placed into scintillation vials, cocktail is added, and the radioactivity is measured in the ^{32}P channel. 100% incorporation of all precursors, with substrate concentrations as specified above, should give 26 μg RNA. Yields generally range from 35 to 45%, but one can use probes even if incorporation is ≈10 to 20%. A transcription reaction should synthesize about 10 μg probe.

3.   Probe purification

Precipitate the riboprobe by adding 10 μl 7.5-M NH$_4$OAc and 60 μl 100% ice-cold EtOH. Precipitate at –20°C, wash, and dry the pellet as usual. The probes are then dried under vacuum and dissolved in 10 mM Tris-HCl buffer (pH 8.0), containing 2 mM EDTA (TE buffer); their concentration should be adjusted to 100 ng/μl. The probes are then denatured by heating to 95°C in a water bath for 3 min before use. The digoxigenin-labeled RNA probes are stored at –20°C.

B.   Hybridization

1.   Dewaxing and postfixation of sections

All solutions used prior to posthybridization washes should be RNase free. It is convenient to prepare l0× stock solutions and to make dilutions as required. Proteinase K and acetylation solutions must be made fresh each time. It is advisable to cover all solutions containing ethanol with plastic wrap to prevent evaporation. Place slides (in a rack) in the following solutions in the order listed:

1. Tissue sections deparaffinized in two changes of xylene, 10 min each
2. Ethanol series, starting at 100% and progressing to 95, 70, 50, and 30%; agitate the slides in each solution until equilibrated (approximately 1 min each)
3. 0.2 N HCl, 10 min
4. 1× PBS, 1 min, three times
5. Slides subjected to proteinase K treatment: 20 ng/ml proteinase K in 50 mM Tris-HCl (pH 7.6), 5 mM EDTA, 10 to 15 min; after this step, the tissue is particularly susceptible to RNases, so use extreme caution
6. 1× PBS, 1 min, three times
7. Postfixation with 4% PFA, 5 min.
8. 1× PBS, 1 min, three times
9. Acetylation: The purpose of this step is to acetylate amino groups in tissues to reduce electrostatic binding of probe.[12] Acetylation also blocks binding of probe to poly-L-lysin-treated slides. These steps must be performed in a hood or another well-ventilated place. Suspend the slide rack above a rapidly rotating stir-bar in 240 ml 0.1-M triethanolamine-HCl (TEA, pH 8.0). Add 600 μl of acetic anhydride; wait 5 min, then add 600 μl more of acetic anhydride. Incubate for a total of 10 min. Care should be taken not to introduce water into the acetic anhydride stock; acetic anhydride would rapidly hydrolyze to acetic acid in the presence of water. Thus, if the agent smells like acetic acid, it should be discarded.

10. Brief wash in 1× PBS three times, dehydration in graded ethanols up to 100%, then air-dried

11. Slides air-dried in a dust-free, RNase-free place, then stored in a sealed container at room temperature; the sections may be stored at this stage in airtight boxes with drying agent at −20°C

2.   Prehybridization

The slides are then prehybridized in closed humid containers at 42 to 48°C with hybridization solution for 1 hr; 100 µl hybridization solution is sufficient for 25 × 75-mm slides with 24 × 50-mm coverslips.

3.   Hybridization

Slides are then incubated with 50 µl/slide hybridization solution containing 20 ng freshly denatured digoxigenin-riboprobe (see above) covered with a coverslip, wrapped in plastic wrap, placed in closed humid containers, and incubated at 42 to 48°C for 4 hr.

4.   Posthybridization washes

1. Remove coverslips in 2× SSC

2. Incubate slides with 40 µg/ml RNase A and 10 U/ml RNase TI in 10 mM Tris (pH 8.0), 1 mM EDTA, 0.5 M NaCl at 37°C for 30 min.

3. Wash in 2× SSC containing 2% normal sheep serum (NSS) and 0.05% Triton X-100 for 1 hr at room temperature with mild agitation.

4. Wash in 0.1× SSC containing 2% NSS and 0.05% Triton X-100 for 30 min at 42°C with mild agitation.

5.   Antibody incubation and washes

To visualize the digoxigenin hapten, a high-affinity antibody, anti-digoxigenin-alkaline phosphate conjugate from Boehringer is used. To prevent nonspecific binding of the antibody, embryos are preblocked with normal sheep serum.

1. Incubate slides in buffer 1 containing 2% NSS and 0.3% Triton X-100 for 30 min at room temperature.

2. Overlay slides with 100 µl diluted antibody conjugate which contains 500 to 1000× dilution of sheep anti-digoxigenin antibody in buffer 1 containing 1% NSS and 0.3% Triton X-100. Cover slides with parafilm and place in a humid chamber. Incubate overnight at 4°C.

3. Put the humidified box back to room temperature for 10 min and then remove the parafilm-covered slides by immersing the slides in buffer 1.

4. Wash twice in buffer 1 for 10 min at room temperature.

6.   Alkaline phosphatase color reaction

1. Wash slides with buffer 2 for 1 to 2 min.

2. Perform the color reaction by incubating the slides in chromogen solution (about 300 μl per slide) in humidified light-tight containers for up to 6 hr, with occasional observation for color development.

3. Stop the color reaction by washing the slides with TE buffer; mount the slides with a coverglass in Aqua mounting medium and allow to dry.

C.   Visualization

The mounted and dried slides are viewed with a microscope under brightfield illumination. Depending on the color reaction used, a brown to dark purple color represents the hybridization signal. If desired, sections can be counterstained prior to mounting with hematoxilin/eosin to reveal nuclei and cytoplasm. Care should be taken not to overstain the sections, since overstaining might occlude the hybridization signal. Images can be obtained using conventional photography with black-and-white or color film. For color films, the necessary filter has to be used to match the light temperature of the light source with the one of the film. A more convenient way is to capture images with a videocamera linked to a computer. The digital images can then be edited with programs such as Adobe Photoshop.

Recipes

Hybridization solution: 50% deionized formamide, 2× standard saline citrate (SSC; 0.15 *M* NaCl, 0.015 *M* sodium citrate), 2× Denhardt's solution (0.02% Ficoll 400, 0.02% polyvinylpyrrolidone, 0.02% bovine serum albumin), 10% dextran sulfate (heat gently and rock to dissolve), 400 μg/ml yeast tRNA (RNase-free), 250 μg/ml salmon sperm DNA, and 20 mM dithiothreitol in DEPC-treated H$_2$O. Make 1-ml aliquots and store at −80°C.

Buffer 1: 0.1 *M* maleic acid/0.15 *M* NaCl, pH 7.5.

Buffer 2: 0.1 *M* Tris, 0.1 *M* NaCl, and 50 mM MgCl$_2$, pH 9.5.

NBT (nitro blue tetrazolium): 75 mg/ml in 75% dimethylformamide.

BCIP (5-bromo-4-chromo-3-indolyphosphate): 50 mg/ml in 100% dimethyl-formamide.

Chromogen solution: 45 μl nitroblue tetrazolium and 35 μl X-phosphate solution in 10 ml buffer 2.

V.   Alternative Labeling Procedures

Instead of ^{35}S-labeled ribonucleotides, one can also use ^{33}P probes.[18] For nonradioactive ISH, fluorescein-tagged probes have been used.[19]

References

1. Koopman, P., in *Essential Developmental Biology, A Practical Approach,* Stern, C.D. and Holland, P.W., Eds., Oxford University Press, Oxford, 1993, p. 233.

2. Holland, P. W., in *Essential Developmental Biology, A Practical Approach,* Stern, C. D. and Holland, P. W., Eds., Oxford University Press, Oxford, 1993, p. 243

3. Wilkinson, D. G., in *Essential Developmental Biology, A Practical Approach,* Stern, C. D. and Holland, P. W., Eds., Oxford University Press, Oxford, 1993, p. 257.

4. Wilkinson, D. G., in *In Situ Hybridization. A Practical Approach,* Wilkinson, D. G., Ed., Oxford University Press, Oxford, 1992.

5. Li, H.-S., Yang, J.-M., Jacobsen, R. D., Pasko, D., and Sundin, O., *Dev. Biol.,* 162, 181, 1994.

6. Albrecht, U., Eichele, G., Helms, J., and Lu, H.-C., in *Molecular and Cellular Methods in Developmental Toxicology.* Daston, G.P., Ed., CRC Press, Boca Raton, FL, 1997, p.23.

7. Reiner, O., Albrecht, U., Gordon, M., Chianese, K. A., Wong, C., Gal-Gerber, O., Sapir, T., Siracusa, L. D., Buchberg, A. M., Caskey, C. T., and Eichele, G., *J. Neurosci.,* 15, 3730, 1995.

8. Parker, S, Eichele, G., Zhang, P., Rawls, A., Sands, A. T., Bradley, A., Olson, E. N., Harper, J. W., and Elledge, S. J., *Science,* 267, 1024, 1995.

9. Lutz, B., Kuratani, S., Cooney, A. J., Wawersik, S., Tsai, S. Y., Eichele, G., and Tsai, M.-J., *Development,* 120, 25, 1994.

10. Lutz, B., Kuratani, S., Rugarli, E. I., Wawersik, S., Wong, C., Bieber, F. R., Ballabio, A., and Eichele, G., *Human Mol. Genet.,* 3, 717, 1994.

11. Xu, X.-C., Clifford, J. L., Hong, W. K., and Lotan, R., *Diag. Mol. Pathol.,* 3, 122, 1994.

12. Xu, X.-C., Ro, J. Y., Lee, J. S., Shin, D. M., Hong, W. K., and Lotan, R., *Cancer Res.,* 54, 3580, 1994.

13. Lotan, R., Xu, X.-C., Lippman, S. M., Ro, J. Y., Lee, J. S., Lee, J. J., and Hong, W. K., *N. Engl. J. Med.,* 332, 1405, 1995.

14. Albrecht, U., Abu-Issa, R., Rätz, B., Hattori, M., Aoki, J., Arai, H., Inoue, K., and Eichele, G., *Dev. Biol.,* 180, 579, 1996.

15. Panoskaltsis-Mortari, A. and Bucy, R. P., *BioTechniques,* 18, 300, 1995.

16. Loda, M., Capodieci, P., Mishra, R., Yao, H., Corless, C., Grigioni, W., Wang, Y., Magi-Galluzi, C., and Stork, P. J. S., *Am. J. Pathol.* 149, 1553, 1996.

17. Krieg, P. A. and Melton, D. A., *Methods Enzymol.,* 155, 397, 1987.

18. Durrant, I., Dacre, B., and Cunningham, M., *Histochem. J.,* 27, 89, 1995.

19. Durrant, I., Brunning, S., Eccleston, L., Chadwick, P., and Cunningham, M., *Histochem. J.,* 27, 94, 1995.

Appendix A. Reagents

Reagent	Manufacturer	Order No.
5'-alpha-S-thio ATP	Boehringer	775746
Acetic anhydride	Sigma	A6404
Anti-DIG antibodies	Boehringer	1087479
Aqua mounting medium	Biomedia Corp. (Foster City, CA)	M02
β-Mercaptoethanol	Sigma	M6250
BCIP	Boehringer	1585002
Canada balsam	Sigma	C1795
DEPC	Sigma	D5758
Developer D19	Kodak	1464593
Dextran sulfate	Sigma	D8906
DIG-UTP	Boehringer	1336 371
DNase 1	Boehringer	776785
Eosin Y	Sigma	E6003
Ficoll 400	Sigma	F2637
Fixer (Kodak)	Sigma	P6557
Formalin	Fisher (Pittsburgh, PA)	SF98-4
Formamide	Boehringer	100 144
Histo Prep	Fisher (Pittsburgh, PA)	SH75-125D
Histoclear	National Diagnostic (Manville, NJ)	HS200
Hoechst 33258	Sigma	B2883
Linbro ("7X")	Flow Labs, Inc.	76-674-93
NBT	Boehringer	1087479
NTB-2 emulsion	Kodak	165 4433
Paraffin, paraplast	VWR Scientific (Denver, CO)	15159-409
Paraformaldehyde	Sigma	P6148
Picric acid	Aldrich (Milwaukee, WI)	25,137-2
Poly A	Sigma	P9403
Polylysine	Sigma	P1399
Polyvinyl pyrrolidone	Sigma	PVP4O
Proteinase K	Boehringer	745 723
RNA transcription kit	Stratagene	200340
RNase T1	Boehringer	109193
RNase inhibitor	Boehringer	799017
RNase A	Boehringer	109 169
^{35}UTP	Amersham	SJ1303
Silica gel desiccant	Sigma	S7625
TESPA	Sigma	A3648
Triethanolamine	Sigma	T1377
Xylene	VWR Scientific (Denver, CO)	EMXX0060-3
Yeast tRNA	Boehringer	109 495

Appendix B. Supplies and Equipment

Item	Manufacturer	Order No.
Coverslips	Corning Inc. (New York, NY)	583330
Cryostat	Bright Instruments (Huntingdon, England)	OTF/AS/EC/MR
Cronex autoradiography film	DuPont	572160
Glass histology racks	VWR Scientific (Denver, CO)	25463-009
Histological water bath	Lab-Line Instruments (Melrose Park, IL)	26103
Mechanical convection oven	Bellco (Vineland, N.J.)	7910-00110
Microscope (bright- and dark-field illumination and fluorescence optics)	Leica, Zeiss, Nikon, Olympus, etc.	
Paraffin microtome	Leica	
Photomicrography setup	Leica, Zeiss, Nikon, Olympus, etc.	
Polypropylene slide mailers	Curtin Matheson (Houston, TX)	362-863
Red safelight	Kodak	152-1525 Serie2
Scintillation vials, glass	VWR Scientific (Denver, CO)	66022-004
Slide warmer	Fisher (Pittsburgh, PA)	12-594
Slide dryer box	Contact geichele@bcm.tmc.edu.	
Slides, Fisher finest	Fisher (Pittsburgh, PA)	12-544-2
Slide holder	Baxter	S7636
Stainless steel racks	VWR Scientific (Denver, CO)	25445-011
Tissue-Tek II staining station	Baxter	S7626-12
Videocamera	Various manufacturers	
Water bath, shaking	Bellco (Vineland, NJ)	7746-12110

Chapter **6**

Synchronization of Cells by Elutriation: Analysis of Cell Cycle-Specific Gene Expression

Steven F. Dowdy, Linda F. VanDyk, and Greg H. Schreiber

Contents

I.   Introduction

A major mechanism that cells use to regulate cell cycle progression is expression of gene products in a cell cycle-dependent fashion.[1,2] The ability to isolate large quantities of cells from specific phases of the cell cycle allows investigation into the cell cycle transcriptional regulation of genes as well as into positive and negative regulation of constitutively expressed genes. Such examples include the transcriptional induction of cyclin A at the G1-to-S phase transition and the induction of cyclin B at the G2-to-M phase transition.[3-6] Other examples include both the activation and inactivation of existing gene products by post-translational modification in a cell cycle-dependent fashion. The retinoblastoma tumor suppressor gene product (pRb) is inactivated by hyperphosphorylation at the late G1 phase restriction point at the hands of cyclin E:cdk2 complexes.[7-9]

There are several methods currently available to synchronize cells from asynchronous populations, including: serum deprivation allowing the isolation of G0 cells,[1,2] amino acid starvation yielding G1 cells,[10] excess thymidine giving rise to cells blocked at the G1-S transition,[11] hydroxyurea blocking cells in early S phase,[12] nocodazole blocking cells at the G2-M phase transition,[13] and centrifugal elutriation yielding cells from all phases of the cell cycle.[14] Cells may be analyzed for gene expression or regulation of protein function by each of these methods. In addition, all of the methods allow for the replating of cell cycle phase-specific cells and can be followed through several subsequent phases of the cell cycle (reviewed in Krek and DeCaprio[15]).

There are caveats to each of these methods to isolate cell cycle phase-specific cells. For instance, serum deprivation leaves cells in a G0 compartment that results in the transcriptional repression of many, if not all, of the cell cycle regulatory genes such as cyclins and cyclin-dependent kinases (cdk), as well as pRb.[7] Therefore, the analysis of the first "G1" involves a G0 to G1 transition and is not necessarily physiologically reflective of the G1 phase in cycling cells such as tumors. Other problems, such as toxicity and/or altered gene-expression patterns, may arise from the use of pharmacological agents to arrest and synchronize cells.

We prefer the method of centrifugal elutriation to isolate cell cycle-specific cellular populations[16] for several reasons. First, this method utilizes exponentially growing cells that are not exposed to pharmacological agents or deprived of serum factors. Second, the cells are elutriated in medium containing serum at 37°C. Third, large quantities of cells from each phase of the cell cycle can be readily obtained for analysis or replating. Fourth, almost all cell types can be cell cycle separated by elutriation. Lastly, the method can be consistently repeated once the elutriation parameters for a given cell type have been established.

The method of centrifugal elutriation involves separating cells based on volume in a specially designed centrifuge rotor (see Figure 6.1A). Briefly, ~1 × 10[9] cells are injected by use of a pump into an elutriation rotor set at a constant speed. The cells are balanced in the elutriation chamber of the rotor by the opposing centrifugal force (proximal to distal) and flow rate from the pump (distal to proximal). The cells become positioned in the elutriation chamber based on their volume and, hence, cell cycle

A.
STANDARD ELUTRIATION CHAMBER

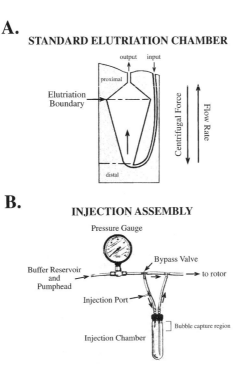

B.
INJECTION ASSEMBLY

FIGURE 6.1

(**A**) The standard elutriation chamber is depicted. Cells enter from the distal port into the chamber and are "floated or balanced" in the chamber by the competition of opposing forces of flow rate and centrifugal force. Small cells move proximal where the flow rate slows due to widening of the chamber, while larger cells remain more distal. The chamber narrows on the proximal side of the elutriation line; as cells cross this line, the flow rate is accelerated and the cells captured. (**B**) The injection assembly is depicted. The volume flow is left to right. The bypass valve is turned to bypass the injection chamber. A syringe containing the cells is attached to the injection port, the valve turned accordingly, and the cells slowly injected into the top of the chamber. The injection port is closed and the bypass turned to allow flow through the chamber; the cells are then drawn into the elutriation rotor. The bypass valve is left in this position for the remainder of the run, as it allows for the capture of bubbles generated by the pump head, thus avoiding entrance into the rotor.

position, such that small G1 phase cells are proximal and larger G2/M phase cells are more distal in the chamber. The small G1 cells are "elutriated" or captured first by increasing the flow rate of the pump and then isolated in an external tube. The process of gradually increasing pump speed is continued until eventually the large G2/M phase cells have been elutriated from the chamber. The entire procedure takes between 1 and 2 hr. The elutriated cells are then analyzed for cell cycle position by DNA content via fluorescence-activated flow cytometry (FACS) analysis, and cells in the specific cell cycle phases of interest are replated (see Figure 6.2). In our hands, cells isolated by centrifugal elutriation can be replated and followed for roughly a complete passage through the cell cycle (see Figure 6.3).

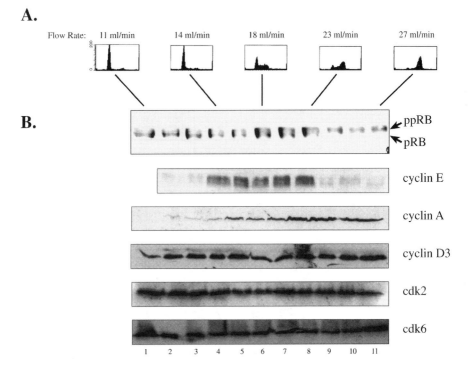

FIGURE 6.2

(A) Typical DNA content FACS analysis of propidium iodide-stained elutriation fraction profiles: The position of cells is primarily comprised of G1 phase cells in the 11-ml/min fraction and G2/M phase cells in the 27-ml/min fraction. (B) Immunoblot analyses of active, hypophosphorylated pRb and slowly migrating inactive hyperphosphorylated ppRb; cyclins E, A, and D3; and cdk2 and cdk6. Hyperphosphorylation of ppRb, as well as expression of cyclins E and A show cell cycle periodicity in cycling cells, while cyclin D3, cdk2, and cdk6 are constitutively expressed throughout the cell cycle.

II.   Materials

A.   Equipment

1.   Elutriation equipment

1. Elutriation rotor, model JE-6B (Beckman) and strobe assembly
2. Elutriation assembly, supplied with JE-6B rotor, including: silanized Tygon tubing, pressure gauge, three-way valves, and tubing Y-connectors
3. Floor model centrifuge, such as Beckman J2 or J6 models
4. Digital pump (Cole-Parmer-MasterFlex #H-07523-20) and pump head (MasterFlex #H-07021-20)
5. Tubing, MasterFlex size #14
6. Ring stand, two

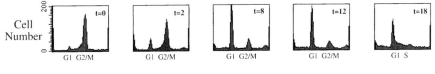

DNA Content

FIGURE 6.3

DNA content FACS analysis of replated elutriated G2/M phase Jurkat T cells: At t = 0 hr, most of the cells are in the G2/M phase; by t = 8 hr, most of the cells have traversed into the G1 phase; and at t = 18 hr, the leading edge of cells has entered the S phase.

7. Water bath

8. 10-ml sterile Luer-Lok syringe

2. Analysis equipment

1. Power pack

2. SDS-PAGE apparatus

3. Semi-dry blotting apparatus (Owl)

B. Reagents

1. Elutriation reagents

1. Tissue-culture medium, such as RPMI, supplemented with 3 to 5% fetal bovine serum and 2× concentrations of antibiotics, such as penicillin and streptomycin

2. Cell dissociation solution (Sigma Chemical)

3. 70 and 30% EtOH solutions

4. 50× propidium iodide (PI) stock: 0.5 mg propidium iodide/ml in PBS(–), stored at 4°C and light protected

5. 40× RNase A stock: 10 mg RNase A/ml PBS(–), stored at –20°C

6. 1× PI working solution: 1× PI stock, 1× RNase A stock, 0.5% NP-40, 1 µg/ml Leupeptin, 50 µg/ml PMSF, 1 µg/ml Aprotinin in PBS(–)

2. Analysis reagents

1. Protein extraction buffer (ELB): 20 mM HEPES (pH 7.2), 250 mM NaCl, 1 mM EDTA, 1 mM DTT, 0.1 to 0.5% NP-40 or Triton X-100, 1 µg/ml Leupeptin, 50 µg/ml PMSF, 1 µg/ml Aprotitin and containing the following phosphatase inhibitors: 0.5 mM NaP$_2$O$_7$, 0.1 mM NaVO$_4$, 5.0 mM NaF

2. 2× sample buffer: 100 mM Tris-HCl (pH 6.8), 200 mM DTT, 4% SDS, 0.2% bromophenol blue, 20% glycerol

3. Protein transfer buffer: 20% methanol, 0.037% SDS, 50 mM Tris, 40 mM glycine
4. Blocking solution: 5% nonfat milk in PBS(–), 0.1% Tween-20
5. Wash buffer: PBS(–)/0.1% Tween-20

III.   Methods

A.   Setup

1. The elutriating centrifuge should be set up per the manufacturer's (Beckman) directions. We (and others) have found that to obtain consistent elutriation of cells it is best that a centrifuge be dedicated as an elutriating centrifuge. This avoids alterations incurred due to the constant removal-placement of the rotor in the centrifuge. For the JE-6B elutriating rotor, either the Beckman J2 or J6 model centrifuges can be used. An area of approximately 2 to 3 linear feet (or more) of bench space next to the centrifuge is required for the pump assembly, etc. The order of placement of equipment is as follows: water bath containing reservoir of elutriating medium, digital pump, ring stand holding the pressure gauge and injecting three-way valve, rotor, sample collecting ring stand. Bench space is also required for storage of collected samples.

2. The pump must be calibrated as per the manufacturer's directions. Use a short segment of MasterFlex tubing #14 and feed from the elutriation medium reservoir through the pump head and then connect to the pressure gauge. MasterFlex tubing must be used for this segment, as other types of tubing will not maintain a consistent linear flow rate with increasing pump head speeds. Attach a segment of clear silanized Tygon tubing on the output side of the pressure gauge. Use this tubing throughout the rest of the assembly. The MasterFlex tubing in the pump head should be changed every other month and the pump recalibrated accordingly.

3. Bubbles are an extremely critical problem in elutriation and must be routinely cleared from the entire assembly prior to use. Initially, pump through water or PBS(–) by bringing the pump rate to 20 to 25 ml/min. Make sure all of the air/bubbles are completely removed from the injection chamber. Tap out bubbles present in the injection assembly and input tubing. After sufficient time has elapsed to fill the entire system, pinch the exit tubing to increase the back pressure to ≤10 psi and hold for several seconds and release. A stream of bubbles should emerge shortly from the rotor. Set the rotor speed to 100 rpm, pinch the exit tubing, and release. Repeat upwards of 8 to l0 times until no bubbles emerge from the rotor. If during the setup or a run the back pressure increases above ~5 psi, a bubble has become lodged or was not initially cleared from the rotor. If cells have already been injected into the rotor, evacuate the cells from the rotor chamber by turning off the centrifuge, collect the cells, centrifuge to reconcentrate, remove the bubble from the rotor chamber as described above, and re-inject the cells.

4. Prior to use, the entire assembly must be sterilized. Sterilize by pumping through ~200 ml of 70% EtOH. Rinse the EtOH out by pumping through approximately 500 ml of sterile PBS(–). Then equilibrate the assembly in elutriation medium present in the 37°C water bath. (This is essentially the medium that the cells to be elutriated grow

in.) The percentage of FBS in the elutriation medium cannot exceed ~3%, or excessive foaming will occur, potentially leading to the introduction of bubbles into the system. Approximately 3 liters of preheated elutriating medium are required. The assembly is now ready to elutriate cells.

5. After elutriation is complete, sterilize the assembly in 70% EtOH and either store in 30% EtOH or dry the entire assembly. Pay special attention to cleaning and sterilizing the three-way valve injection port. Wash out the 70% EtOH with water. We routinely store the entire assembly in 30% EtOH and have not had a problem with subsequent bacterial contamination of the system.

B.   Preparing Cells for Injection into Rotor

1. Prior to use, pre-warm the centrifuge by adjusting the temperature to 30°C and set the rotor speed to 2400 rpm and the pump speed to 8 ml/min. Collect the eluate from the exit tubing in a large container and discard as required.

2. The standard elutriating chamber requires between 5×10^8 and 2×10^9 cells to achieve the separation based on size and, hence, cell cycle. If you are elutriating larger cells, such as HeLa cells, use a lower number of cells on this scale; if the cells of interest are small, such as lymphocytes, use the higher number of cells. We routinely elutriate 8×10^8 Jurkat T cells.

3. Cells must be single, unclumped cells, whether the source was from nonadherent or adherent cells. Concentrate the cells (after removal from dishes, if required) by centrifugation and resuspend in 10 ml cell dissociation solution. This will further ensure that the population is separated into single cells. Draw 10 ml of cells into a 10-ml sterile Luer-Lok syringe and attach to the closed three-way valve on the injection assembly (Figure 6.1B).

4. Bypass the injection chamber by turning the bypass valve accordingly. Open the port to which the syringe is attached on the three-way valve and slowly inject the cells into the injection chamber. After this is completed, close the injection valve port and turn the bypass to allow the elutriation medium to enter the injection chamber. The cells will slowly be drawn into the spinning rotor. This process takes about 10 min. Save a small amount of cells as the initial starting population for FACS analysis (see Section III.D.1). Others have opted to exclude the injection chamber assembly; however, the pump head continually generates a stream of bubbles due to the presence of 3% serum in the elutriation medium, and the injection chamber serves as a capture reservoir.

5. Allow an additional 10 min for the cells to reach a size equilibrium within the rotor. The cells may be viewed by use of the strobe light assembly; however, this ability is not required for good elutriation of cells.

C.   Elutriating Cells from the Rotor

1. To elutriate cells from the rotor, the pump flow rate is slowly increased and the elutriated cells collected. The first fractions usually contain any dead cells present in the

population and are quickly followed by G1, S, G2, and M phase cells. The purity of each subsequent fraction is entirely dependent on the elutriation of the prior size cells from the elutriation chamber.

2. Slowly lower the speed of the rotor to 2000 rpm at a rate of a 100-rpm decrease per 1 min (see below for variations of rotor speed and pump flow rates). After achieving a constant rotor speed of 2000 rpm, the rotor speed is not adjusted. The Beckman manual suggests an alternative elutriation approach of continually decreasing the rotor speed; however, this method is more inaccurate when compared to increasing the flow rate of the digital pump and, therefore, we do not recommend it.

3. Increase the flow rate from 8 to 9 ml/min and collect 100 ml of elutriated cells from the exit tubing in two sterile 50-ml disposable tubes. This is best done by placing a rack of labeled 50-ml tubes under the collection exit tube and sliding it over to collect each new fraction. If the elutriated cells are to be replated, either leave the tubes at room temperature or place at 37°C. If the cells are to be analyzed directly for expressed genes or post-translational protein modification, place tubes into an ice bucket.

4. Increase the flow rate from 9 to 10 ml/min and collect 100 ml as described above. Continue to increase the flow rate from one fraction to the next by 1 ml/min while collecting 100 ml fractions until a rate of 21 ml/min has been achieved.

5. After 21 ml/min, the flow rate is increased by 2 ml/min for each subsequent fraction up to 29 ml/min, and 100 ml of cells are collected.

6. After 29 ml/min, the flow rate is increased by 3 ml/min until a flow rate of 38 ml/min has been achieved and 100 ml fractions are collected.

7. The cells remaining in the elutriation chamber usually consist of clumps and/or tetraploid cells in late G2/M phase. These cells are vacated from the chamber by turning off the centrifuge while keeping the pump flow rate at 38 ml/min. The entire elutriation usually takes 1 to 2 hr to complete.

8. The assembly is cleaned as described above (Section III.A). Check the centrifuge chamber for leaks; if any solution is on the centrifuge walls then the rotor should be disassembled according to the manufacturer's directions.

D. Analysis of Elutriation Fractions

1. Analysis of cell cycle position and replating

1. The elutriated fractions are analyzed for cell cycle position by quantitative propidium iodide DNA staining and FACS. Invert each tube to mix the cells, take 5 ml from both 50-ml tubes for a given fraction, combine, place into a 15-ml tube, and pellet the cells by centrifugation. Aspirate the supernatant and gently resuspend the cell pellet in 200-µl of ice-cold PBS(−). Slowly mix in 800 µl of ice-cold 100% EtOH while gently rotating the tube. Place the tubes on a rotating wheel at 4°C for 10 min. Centrifuge the cells at 1800 rpm for 5 min at 4°C and aspirate the supernatant. Resuspend the cells in 500 µl of 1× PI staining solution and analyze on a FACS. Alternatively, the cells may be resuspended after the initial spin in 0.5 ml of 1× PI staining solution containing 0.5% NP-40 and protease inhibitors, placed on ice, and then analyzed by FACS within the next several hours. We prefer this method due a decreased time commitment.

2. The PI-stained cells are analyzed for cell cycle position by FACS, utilizing the FL-2 detector by setting up a dot matrix window of FL-2 width on the x-axis and FL-2 area on the y-axis (Figure 6.2A).

3. Count the cells present in each fraction by use of a hemacytometer or Coulter counter. When starting with 8×10^8 cells, we usually obtain between 5×10^6 and 5×10^7 cells per fraction. The large variation in cell numbers will depend on several factors, including phase of interest and cell cycle profile of the cells prior to elutriation.

4. After determining by FACS analysis the cell cycle profile of each elutriated fraction, cells from the fraction of interest can be replated and followed through subsequent phases of the cell cycle. Centrifuge the fraction(s) of interest present in the 50-ml tubes at 1800 rpm for 10 min, aspirate the supernatant, and place in culture at the required density in fresh medium. In our hands, Jurkat T cells can be followed through an entire cell cycle after replating at 1×10^6 cells/ml (Figure 6.3).

2.   Immunoblot analysis

1. Centrifuge the identical number of cells from each elutriated fraction at 1800 rpm for 5 to 10 min at 4°C. Aspirate the supernatant, wash cells in 10 ml ice-cold PBS(–), centrifuge, and aspirate as above.

2. Resuspend the cell pellet in 0.2 to 1 ml ice-cold extraction buffer (ELB[17]), transfer to a 1.5-ml Eppendorf tube, and place on ice for 30 min with occasional mild inverting. Spin out insoluble particulate matter from the cellular lysate in a microfuge at 12 K, 4°C, for 10 min. Transfer the supernatant to a new Eppendorf tube. The amount of ELB required will vary depending on the number of cells in the pellet and on whether the cellular lysates are to be analyzed by immunoblotting analysis (see below) or are to be labeled with ^{35}S-methionine or ^{32}P-orthophosphate and immunoprecipitated (use 1 ml).

3. For immunoblot analysis, add 60 µl 4× sample buffer and boil the sample for ~5 min, centrifuge 12 K for 10 sec, cool tube on ice, and load lysate onto SDS-PAGE.[18]

4. After running the SDS-PAGE, separate the glass plates and trim-down the gel with a razor blade by removing the stacking gel and any excess on the sides or bottom. Measure the trimmed gel size and cut six sheets of 3MM Whatman filter paper and one sheet of nitrocellulose (NC) filter to the same size.

5. Soak two sheets of cut 3MM filter paper in transfer buffer and place on semi-dry transfer unit. This can be done by filling a flat container with transfer buffer and dipping the 3MM paper into it. Place one soaked 3MM sheet on the back of gel and rub slightly to adhere to gel. Then invert the glass plate with the gel/MM still stuck to it and peel the gel/3MM away from the glass with the use of a razor blade. Place it 3MM-side-down onto the soaked 3MM sheets on the semi-dry unit. Soak the cut NC filter with transfer buffer and place it on top of gel followed by three presoaked 3MM sheets on top of the filter. Gently squeegee out bubbles and excess buffer with the back of your little finger or by rolling a small pipette over the stack. It is important to squeegee out the bubbles, but avoid excess squeegeeing that would result in drying the stack. Mop up the excess buffer on the sides of the stack present on the transfer unit. Place the top on the transfer unit. In this configuration, the bottom plate is the cathode (negative) and the top is the anode (positive).

6. Transfer the proteins to the NC filter at constant 10 V for 1.5 to 2 hr. The starting current will vary from ~80 to 300 mAmp depending on the surface area of the gel and will drop to around 80 to 140 mAmp by the end of the transfer.

7. Disassemble the transfer apparatus and place the NC filter, protein side up, into a small flat container containing ~50 ml of blocking solution (5% nonfat milk in PBS(–)/0.1% Tween-20). Place on a rotating shaker platform at 50 to 80 rpm at room temperature (RT) for 30 to 60 min.

8. Place NC filter into a "seal-a-meal" bag; add 1 ml of diluted primary antibody in 2.5% nonfat milk/PBS(–)/Tween-20; and shake at 50 to 80 rpm at RT for 1 hr to overnight. The dilution of the primary antibody will vary depending on the antibody avidity[17] for the immobilized epitope. We routinely dilute the primary antibody 1:1000–5000.

9. Cut open a corner of the bag and drain the primary antibody solution. Rinse the filter twice with ~50 ml blocking solution and remove the filter from the bag, place in a flat container containing ~200 ml wash buffer (PBS(–)/Tween-20) and shake at RT for 15 min.

10. Drain the wash buffer, and repeat step 9 two more times.

11. Place the filter into a new "seal-a-meal" bag, add 0.5 to 1 ml of conjugated anti-primary secondary antibody, such as horseradish peroxidase-conjugated secondary antibodies, diluted 1:1000; seal; and place on shaker at RT at 50 to 80 rpm for 1 hr.

12. Rinse and wash as described above in steps 9 and 10.

13. Place the washed filter on a slab of glass, such as a gel plate; add 1 ml of mixed enhanced chemiluminescence (ECL) reagent, incubate 1 min, and dry filter by placing on a sheet of 3MM Whatman paper. Place autoradiographic (ARG) film over developed filter and expose for 5 sec to 5 min.

3.   Immunoprecipitation analysis

1. For immunoprecipitation analysis, replate the appropriate number of cells in labeling medium, such as ^{35}S-methionine or ^{32}P-orthophosphate, for 3 to 5 hr. Centrifuge the labeled cells at 1800 rpm, 4°C, for 5 min. Aspirate the supernatant, wash in 10 ml ice-cold PBS(–), repeat spin.

2. Resuspend the cell pellet in 1 ml ice-cold ELB and treat as described above (Section III.E, step 2).

3. Pre-clear lysate by the addition of 50 μl killed *Staphylococcus aureus* cells, cap tube, and place on rotating wheel at 4°C for 30 to 60 min. Remove *S. aureus* cells by centrifugation of lysate at 12 K, 4°C, for 10 min. Transfer lysate to fresh Eppendorf tube, making sure to leave the last 50 μl or so of lysate behind with the pellet. The presence of contaminating *S. aureus* cells in this portion of the sample will reduce the amount of immunocomplexes recovered if included.

4. Add primary antibody to the pre-cleared lysate supernatant, usually 100 to 200 ml in case of hybridoma supernatants, and 3 to 5 μl of commercially purified antibodies. Add 30 μl protein A agarose beads, cap the tube, and place it on a wheel at 4°C for 2 to 4 hr or overnight. If primary mouse antibody isotype is IgGl or unknown, add 1 μl rabbit anti-mouse IgG to allow indirect binding of the primary antibodies to the protein A agarose.[17]

5. After incubation with primary antibody, perform a "1-*g* spin" by placing the tube on ice for ~15 min; aspirate the supernatant to just above the protein A agarose bed level. Stop aspirating at the top of the agarose beads, avoiding drying the beads. Wash the agarose beads by the addition of 1 ml ice-cold ELB, cap and invert tube several times, and centrifuge at 12 K, 4°C, 30 sec. Aspirate supernatant and repeat two to three more times. Again, avoid drying the beads.

6. After the final 30-sec spin, aspirate supernatant off the protein A beads until just dry, add 30 µl of 2× sample buffer. Boil sample for ~5 min, centrifuge at 12 K for 10 sec, cool tube on ice, and load immunocomplexes onto SDS-PAGE.

7. After running the SDS-PAGE, dry the gel and expose to ARG film or phosphorimager screen.

IV.   Conclusions

The method of obtaining synchronized cells by elutriation offers many advantages over other methods of cell cycle synchronization as outlined above; however, the investigator must be willing to invest in a substantial amount of dedicated equipment. All types of cells, both adherent and suspension cells, can be elutriated. Once the protocol for a given cell type has been determined, it is fixed and does not need to be re-established in the future, provided the identical parameters are always employed. The parameters for successful elutriation, including rotor speed, number of cells injected, flow rate changes, and volumes collected, will vary greatly from one cell type to another. The parameters given above should be used as a starting point and fine tuned for specific cell types. If the investigator is focusing on G1 or G2/M phase cells, the protocol can be dramatically shortened. As an example, if G2/M phase cells are desired, jump ahead to approximately 17 ml/min and collect ~300 ml of media/cells.

The use of synchronized cells isolated by the method of elutriation allows investigation into cell cycle-specific phases from unaltered cycling cells, such as occur in tumors, and can yield profound differences in expression patterns from other methods employed, such as the commonly used serum-deprivation protocol. As an example, using serum deprivation, D-type cyclins were initially categorized as having a late G1 expression pattern.[19] This led to a hypothesis that D-type cyclins in concert with their cyclin-dependent kinase (cdk) catalytic subunits, cdk4 and cdk6, performed the initial inactivating hyperphosphorylation of pRb at the late G1 restriction point. However, analysis of D-type cyclin expression from elutriated cycling cells demonstrates a constitutive cell cycle pattern of expression (Figure 6.2B), essentially placing this view in conflict. In contrast, the initial expression of cyclin E in cycling cells closely correlates with the appearance of the more slowly migrating, inactive hyperphosphorylated pRb forms (designated ppRb in Figure 6.2B). Thus, the use of unperturbed synchronized cells from an elutriation protocol can have a profound impact on previously observed differences utilizing methods that alter the physiology of the cell.

Acknowledgments

We thank Drs. J. A. DeCaprio for protocols and H. Piwnica-Worms for helpful suggestions. L. F. VD. was supported by a postdoctoral fellowship from the ACS (#PF-4379). S. F. D. is an Assistant Investigator of the Howard Hughes Medical Institute.

References

1. Pardee, A. B., *Science,* 246, 603, 1989.
2. Rollins, B. J. and Stiles, C. D., *Adv. Cancer Res.,* 53, 1, 1989.
3. Solomon, M. J., *Curr. Biol.,* 5, 180, 1993.
4. Sherr, C. J. and Roberts, J. M., *Genes Dev.,* 9, 1149, 1995.
5. Weinberg, R. A., *Cell,* 81, 323, 1995.
6. Lees, E., Faha, B., Dulic, V., Reed, S. I., and Harlow, E., *Genes Dev.,* 6, 1874, 1992.
7. Mittnacht, S. and Weinberg, R. A., *Cell,* 65, 381, 1991.
8. Hinds, P. W., Mittnacht, S., Dulic, V., Arnold, A., Reed, S. I., and Weinberg, R. A., *Cell,* 70, 993, 1992.
9. Lees, J. A., Buchkovich, K. J., Marshak, D. R., Anderson, C. W., and Harlow, E., *EMBO J.,* 10, 4279, 1991.
10. DeCaprio, J. A., Ludlow, J. W., Lynch, D., Furukawa, Y., Griffin, J., Piwnica-Worms, H., Huang, C., and Livingston, D. M., *Cell,* 58, 1085, 1989.
11. Bootsma, D., Budke, L., and Vos, O., *Exp.Cell Res.,* 33, 301–309, 1964.
12. Hubermann, J. A., *Cell,* 23, 647, 1981.
13. Zieve, G. W., Turnbull, D., Mullins, J. M., and McIntosh, J. R., *Exp. Cell Res.,* 126, 397, 1980.
14. Grabske, R. J., *Fractions,* 1, 1, 1978.
15. Krek, W. and DeCaprio, J. A., *Methods Enzymol.,* 254, 114, 1995.
16. Draetta, G. and Beach, D., *Cell,* 54, 17, 1989.
17. Harlow, E. and Lane, D., *Antibodies: A Laboratory Manual,* Cold Spring Harbor Laboratory Press, Cold Spring Harbor, New York, 1988.
18. Judd, R. C., *Methods Mol. Biol.,* 32, 49, 1994.
19. Sherr, C. J., *Science,* 274, 1672, 1996.

Programmed Cell Death;
Cell Culture

Chapter **7**

Molecular Analysis of Programmed Cell Death in Mammals

Rajagopal Gururajan, Jill M. Lahti,
and Vincent J. Kidd

Contents

0-8493-4411-5/98/$0.00+$.50

I.   Introduction

The term "apoptosis" was originally used by Kerr and Wylie to distinguish a form
of cellular death that required the active participation of the dying cell, a process that
was quite distinct from cellular necrosis.[1,2] Their studies demonstrated that an
orderly and defined series of biochemical, molecular, and cellular alterations accompanied such targeted cell death. Programmed cell death (PCD) more appropriately refers to the apoptotic death of specific cells during development and, therefore, is not necessarily the functional equivalent of cellular apoptosis. Studies from
a number of laboratories have demonstrated that many of the gene products involved
in PCD are identical, or closely related, to those responsible for apoptosis.[3,4] These
similarities have blurred any differences between these two processes, often resulting in the interchangeable use of these two terms. Whether new mRNA and protein
expression are required for the proper execution of the apoptotic program has been
a subject of considerable debate. Experimental evidence demonstrating that vertebrate cells can undergo apoptosis in the absence of a nucleus and/or new mRNA/
protein synthesis suggests that the nuclear events may not be relevant to these forms
of apoptosis;[5-9] however, death of interleukin-2 (IL-2)-dependent T-cells and neuronal cells from avian embryos is accompanied by the expression of new mRNAs
and proteins. Inhibitors of RNA and/or protein expression have also been used to
block apoptosis effectively. Obviously, while some cells may have to express, *de
novo*, proteins necessary for their own execution, others may already possess the
death machinery in an inactive form that can be rapidly activated in response to the
appropriate signal(s). It has been proposed that the activation of an apoptotic
signaling cascade is normally suppressed by survival signals. Withdrawal of these
signals and/or the activation of cell death receptors can rapidly initiate apoptosis.[10,11]

Morphological changes in apoptotic cells, such as active membrane blebbing
and the collapse of rapidly condensing chromatin against the nuclear periphery, are
still commonly used to identify apoptotic cells. These morphological changes are
often seen in response to a variety of apoptotic stimuli in many different cell types.
Subsequently, a number of biochemical and molecular features of apoptotic cells
have been defined which are also very similar to those associated with PCD during
normal development.[11] In fact, Horvitz and coworkers[3,12] have shown that a group of
genes function to regulate the death of specific cells during development of the
nematode, *Caenorhabditis elegans.* By genetically manipulating *C. elegans,* several
genes that are required for both lineage-specific and nonlineage-specific cell death
in nematodes have been isolated. The cell death abnormal (*ced*) genes of *C. elegans,*
ced-3 (an ICE-related protease) and *ced-4,* are required for cell death, while the *ced-
9* gene encodes a bcl-2 homologue that suppresses cell death.[13] The precisely timed

death of specific cells during *C. elegans* development is mediated by the coordinated action of these *ced* genes, as well as two additional genes that suppress cell death when expressed (the cell death specification, or *ces* genes[14]); hence, the term "programmed cell death" was originally used to define a process of cell death intimately linked to the proper temporal and spatial death of cells during development. Another seven *C. elegans* genes are required for proper phagocytosis of the cellular corpse, while the *nuc-1* gene facilitates the digestion of genomic DNA from these dead cells.[15]

The products of many of these genes have been functionally conserved in higher vertebrates, and they constitute a growing family of effectors that are essential for mammalian apoptosis.[11] During developmentally regulated PCD, various physiological stimuli (e.g., hormones) are responsible for regulating the expression of the *ces/ced* genes. However, vertebrate cells undergoing apoptosis are often responding to death effectors that are triggered by internal or external stimuli unrelated to development, including changes in cellular environment (e.g., the presence or absence of survival factors), unscheduled/scheduled changes in DNA replication and/or chromosome division that occur during the maturation and/or aging of cells, or environmental stress (e.g., radiation).[11]

The criteria and techniques used to establish whether vertebrate cells are undergoing apoptosis have evolved substantially during the past 5 years. There has been a renewed interest in understanding the process(es) that control apoptosis and PCD, as well as determining how these events are functionally linked to the regulation of other cellular processes, such as the cell cycle.[11,16] Mammalian and other vertebrate cell cycles are exquisitely controlled by a complex network of positive and negative signals/regulators, which include cyclins, cyclin-dependent protein kinases (CDKs), and cyclin-dependent protein kinase inhibitors (CKIs).[17,18] These proteins and the genes that encode them respond to many diverse environmental cues, including cellular growth factors and genetic damage. Eukaryotic cells have developed mechanism(s) to monitor the fidelity of their progression through the cell cycle (commonly referred to as cell cycle "checkpoints") and to determine whether all of the required events (e.g., replication of DNA, duplication of chromosomes) have been faithfully completed before the cell divides.[19,20] If extensive cellular damage occurs, these cells can trigger their own death (cellular suicide), thereby eliminating "abnormal" cells before any harm can be done to the organism. Elimination of such abnormal cells may occur throughout cell cycle progression.[21] For example, apoptosis may be triggered by DNA damage during the Gl phase, but cell death may be delayed until the G2/M-phase or after entry into the next cell cycle. Many cells (e.g., lymphocytes) depend upon survival factors that they, or a neighboring cell, express to avoid cellular suicide or, when appropriately triggered, to induce apoptosis;[22] therefore, migration of a cell into an inappropriate environment in which the survival factor is not expressed can trigger apoptosis of these maverick cells to prevent systemic dysfunction(s) that might occur as a result. Maintaining the proper balance between cellular proliferation, death, and migration is crucial for homeostasis. Furthermore, we now know that defects in apoptotic signaling pathways contribute significantly to the development of heritable genetic disease, autoimmune disorders, and cancer.[23-25]

II. Cellular and Molecular Features that Are Commonly Associated with Apoptosis, Programmed Cell Death, and Necrosis

A. Morphological Changes

The morphological changes associated with cells undergoing apoptosis or PCD are quite distinct from those associated with necrosis.[11] For instance, apoptotic cells tend to shrink and undergo membrane blebbing, and their chromosomes rapidly condense and aggregate around the nuclear periphery. In contrast, necrotic cells swell and burst, and the resulting cellular debris is removed by rapidly recruited macrophages. Electron microscopy of apoptotic cells has revealed that the margination and condensation of the chromatin, prominent vacuolization of the endoplasmic reticulum, cell fragmentation, and autophagocytosis (characterized by vacuoles seen within these dying cells) are unique to these cells. Although these morphological differences have been the basis for defining apoptosis, it is now clear that not all of these morphological changes apply universally to all apoptotic cells.[6] There are often distinct differences in the appearance of apoptotic cells, as well as differences in biochemical and molecular processes that constitute apoptotic signaling pathways. For instance, microscopic analysis of Fas-induced apoptosis in three different human T-cell lines shows their markedly different cellular morphologies (Figure 7.1). The CEM-C7 T cells shrink dramatically, and once membrane blebbing begins these cells "spit off" small apoptotic bodies which often contain a small portion of the condensed chromatin debris. In contrast, Jurkat T cells undergoing Fas-mediated apoptosis shrink, but they do not generate the apoptotic bodies. Finally, Molt-4 T cells shrink and often appear as clusters, but much like Jurkat cells they do not form apoptotic bodies.

B. ICE-Like (Caspase) Proteases

In the past year, a large cadre of cysteine proteases, related to the interleukin 1-β-converting enzyme (ICE) have been identified as essential components of many apoptotic signaling pathways. The *C. elegans ced-3* gene was isolated, and its predicted protein sequence demonstrated that it was very similar to the mammalian ICE protein.[26] Prior to the identification of *ced-3* as a homologue of ICE, the role of proteases in apoptotic pathways was based on protease inhibitor studies that were often inconsistent due to their broad, nonspecific nature.[27] Even though the ICE-related proteases are, functionally, cysteine proteases, their sequence is more homologous to some serine proteases; this may help to explain why serine protease inhibitors, such as N-tosyl-L-phenylalanylchloromethyl ketone (TPCK), block apoptotic signals.[28] Following the identification of the *ced-3* gene as an ICE-like protease, ten additional ICE-like proteases (now referred to as Caspase 1–10) were isolated by polymerase chain reaction (PCR)-based methodologies.[13,29] Many of these Caspase proteases induce apoptosis when expressed ectopically in mammalian

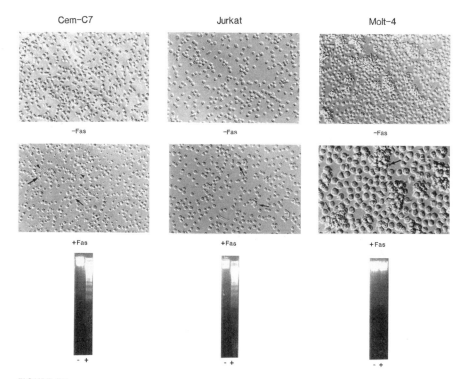

FIGURE 7.1

A comparison of cellular morphologies and DNA integrity in three different human T-cell lines undergoing Fas-mediated apoptosis: The human T-cell lines CEM-C7, Jurkat, and Molt-4 were cultured in the absence (–Fas) and presence (+Fas) of Fas mAb for ~4 hr. Apoptotic cells are indicated by the arrows, and apoptotic bodies are denoted by the arrowheads. DNA isolated from these cells was analyzed by agarose gel electrophoresis and the DNA visualized by staining with ethidium bromide. Apoptotic DNA ladders are evident in both the CEM-C7 and Jurkat DNAs isolated from cells treated with the Fas mAb (+), but not in the untreated samples (–) or in either DNA sample from the Molt-4 cell line.

cells; however, it should be pointed out that ectopic expression of some serine proteases, with no known apoptotic function, can also induce cell death.

The active Caspase enzymes usually consist of a variable aminoterminal pro-domain and two enzyme subunits (an ~17-kDa subunit containing the active site of the enzyme, as well as an ~12-kDa subunit) which are cleaved from the proenzyme and often heterodimerize.[13] The functional Caspase enzyme is generated by the action of other Caspase proteases or granzymes during apoptosis.[11,30] The activation of the Caspase proteases is linked to the aggregation of cell surface receptors (including the tumor necrosis factor [TNF], CD95 [Fas], CD40, and nerve growth factor [NGF] receptors), which occurs when the appropriate ligand is expressed and bound. These receptors lack cytoplasmic protein kinase domains, which are often associated with mitogenic receptors, but they all contain a region of identity within the cytoplasmic portion of the receptor, known as the "death domain" (DD; Figure

7.2). The DD has been implicated in the recruitment of specific proteins that interact with these receptors and potentiate either a death signal or a proliferative signal.[31] For example, once the TNF and Fas receptors are activated by binding their respective ligands, cytoplasmic proteins that contain similar DDs (e.g., TRADD, FADD/MORT1, and FLICE/MACH1) are recruited from the cytoplasm and interact with the DD of these receptors, as well as each other (Figure 7.2). In some cases (e.g., FLICE/MACH 1), the predicted peptide sequence of the carboxyl terminal region of the protein following the death domain resembles the peptide sequence of the ICE-like proteases.[32] Thus, a proteolytic cascade linked to these receptors may be an integral part of many apoptotic pathways. It should also be noted that these same TNF-related receptors (particularly tumor necrosis factor receptor [TNFR] and FasR) can also function in an anti-apoptotic capacity, by recruiting distinct cytoplasmic proteins (e.g., TNFR-associated factors [TRAF]) that also contain death domains. These interactions often result in the activation and nuclear translocation of NF-κB.[33]

C.   Premeditated Death
and the Developing Fly

In contrast to the vertebrate TNFR-related receptors described above, the fly *Drosophila melanogaster* expresses a group of unique cytoplasmic DD-containing proteins from several developmentally regulated and linked genes — *reaper (rpr), head involution defective (hid),* and *grim* — and all are essential for triggering PCD.[34] Much like the TNFR, FasR, NGFR, FADD/MORT1, TRADD, and FLICE/MACH1 proteins in vertebrates, these proteins contain death domains; however, none of these proteins contains Caspase protease domains, and in the case of the 65 amino acid *rpr* gene product, they encode very little else other than a DD. Even though the exact function of this death domain, such as that contained in *reaper*, is not yet known, its function may be linked to promoting protein:protein interactions. In a recent study of the *reaper (rpr)* gene, it was shown that mutations in its DD, which mimic mutations in the DDs of TRADD, FADD/MORT1, and FLICE/MACH1, affected their ability to bind to the TNFR and FasR but did not influence the ability of reaper to induce cell death per se.[35] However, the baculovirus p35 and *iap* gene product (discussed below) has been shown to block the *rpr* death signal effectively by inhibiting Caspase protease activity.[36] Loss-of-function mutations in genes such as *rpr* in *D. melanogaster* and *ced-3* in *C. elegans,* prevent the normal death of specific cells during development.[26,34] In mammals, the elimination of either the *ICE (Caspase 1)* or *CPP32 (Caspase 3)* genes in mice by homologous recombination produces two very distinctive phenotypes.[37-39] Deletion of the *ICE* gene did not affect the normal development of these mice or the ability of their thymocytes to respond properly to glucocorticoids or UV; however, thymic cells from these ICE–/– mice were insensitive to death induced by the Fas ligand. Interestingly, deletion of the *CPP32 (Caspase 3)* gene in mice produces developmental defects that primarily affect the brain, much like the neuronal hyperplasia that occurs in *C. elegans* when the *ced-3* gene is disrupted.[39] Once the function of the various Caspase proteases, as well as other death effectors,

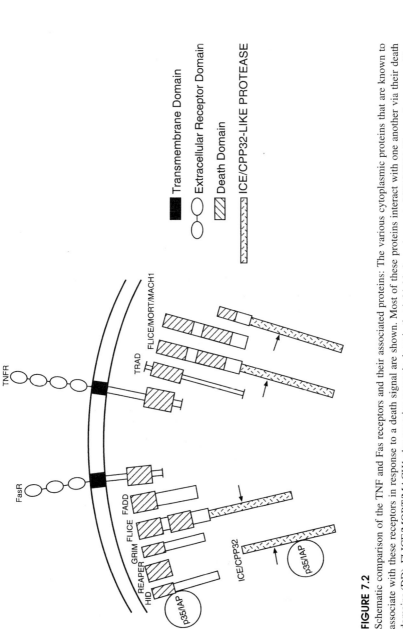

FIGURE 7.2

Schematic comparison of the TNF and Fas receptors and their associated proteins: The various cytoplasmic proteins that are known to associate with these receptors in response to a death signal are shown. Most of these proteins interact with one another via their death domains (DD). FLICE/MORT/MACH1 also contains a protein domain that is highly homologous to ICE/CPP32 (Caspase) proteases. The arrow indicates the cleavage site within the protease that is used to generate the two enzymatically active subunits. The baculovirus p35 inhibitor of apopain proteases (IAP) is also shown.

is elucidated by molecular and genetic analyses, the similarities and differences of these pathways should become more clear.

D. bcl-2 and Its Related Family Members; Death Antagonists and Agonists

The *bcl-2* gene was initially described as the proto-oncogene involved in a t(14;18) translocation that occurs in human follicular lymphomas.[40-42] When bcl-2 is overexpressed it delays, but does not completely block, apoptosis induced by a variety of signals.[16] The ability of increased levels of bcl-2 to protect cells from death suggested that this protein negatively regulated essential apoptotic genes. For the first time, a gene whose function was linked to the regulation of cell death was implicated in oncogenesis, demonstrating that cancer could be viewed as a disease associated with alterations in the ability of the cells to die as well as proliferate. A number of elegant studies from several laboratories have shown that bcl-2 can protect cells from apoptosis linked to the Fas receptor,[43] trophic factor withdrawal (e.g., neurons[44]), *c-myc* overexpression,[45] and growth factor-mediated programmed cell death.[46] Just how bcl-2 protects cells from death is not yet entirely clear. Its localization to various cytoplasmic membranes, as well as the recently derived crystal structure of a related family member, bcl-x_L (demonstrating that the arrangement of α-helices within bcl-x_L resemble the membrane translocation domain of diphtheria toxin), suggests that these proteins could be involved in regulating cellular homeostasis by functioning as membrane pores.[47] A number of bcl-2-related proteins have been identified, including bcl-x_L, bcl-x_S, bax, bak, and bad.[16] Unlike bcl-2, many of these related proteins, including bax, bak, bcl-x_S, and bad, function as agonists of programmed cell death. Many of these bcl-2-related proteins can either homo- or heterodimerize with each other. Experiments have shown that the ratio of death antagonists (bcl-2, bcl-x_L, mcl-1) to agonists (bcl-x_L, bax, bak) within a cell influences its fate;[16] this ratio varies in different cell types, and the expression of one or more of these bcl-2 family members is frequently altered in response to certain apoptotic stimuli. Overexpression of bcl-2, or other agonists, will effectively block Caspase activity induced by a variety of apoptotic signals.[43] Thus, the endogenous levels of these proteins in a given cell should, at least in part, reflect their sensitivity to apoptotic signals.

E. Viral and Peptide Inhibitors of Apoptosis

The identification of proteases as downstream effectors of TNFR-related apoptotic pathway(s) has significantly contributed to the discovery of both endogenous and synthetic inhibitors of these proteins. Studies of viral genes that prevent host cell death to allow replication of its viral genome led to the discovery of viral genes that inhibit apoptosis.[48,49] These viral genes include *crmA* from the cowpox virus and the p35-kDa and inhibitor of apoptosis (IAP) proteins found in baculovirus. The cowpox virus-encoded cytokine response modifier A (crmA) is a 38-kDa protein that effectively blocks many apoptotic signals, apparently by inhibiting ICE activity.[50] The lAP proteins effectively inhibit apoptosis initiated by a wide variety of stimuli through its ability to irreversibly bind to the active site of the Caspase proteases, thereby preventing the normal proteolytic processing of cellular substrates (e.g., poly[ADP]-ribose polymerase [PARP], U1-70K, PKCδ, etc.[48,51,52] *D. melanogaster* and human homologues of the baculovirus *iap* gene have now been isolated, suggesting that these proteins are critical components of cellular apoptotic signaling pathway(s), the function of which has been subverted by many viruses.[53]

In an attempt to understand the mechanism(s) involved in apoptotic signaling pathway(s), synthetic peptide inhibitors that mimic the cleavage sites of Caspase proteases (usually at an Asp-X bond) were chemically derived.[7,54,55] These peptides, normally 3 to 4 amino acids in length, include zVAD-fmk (benzyloxycarbonyl-Val-Ala-Asp-fluoromethylketone), DEVD-cmk (acetyl-Asp-Glu-Val-Asp-chloromethylketone), and YVAD-cmk (acetyl-Tyr-Val-Ala-Asp-chloromethylketone). Even though these peptides are related to one another, subtle differences in their sequence dramatically affect the specificity of their substrate inhibition. The zVAD-fmk and YVAD-cmk peptides are capable of effectively blocking most apoptotic stimuli (e.g., Caspases 1 and 3), while DEVD-fmk primarily inhibits Caspase 3 (CPP32-like) protease activity.[7,43] These peptide inhibitors are generally soluble, relatively stable (a half-life of 4 to 5 hr), and inhibit several different death stimuli in a dose-dependent manner (e.g., Fas-induced death is maximally inhibited by 10 mM zVAD-fmk[7]). In fact, the rate of enzyme inactivation by these peptide inhibitors can be determined by addition of excess inhibitor to the enzyme reaction and observing the nature of the time-dependent inhibition of substrate hydrolysis.[43] These peptide substrates include YVAD-AMC (acetyl-Tyr-Val-Ala-Asp-aminomethylcoumarin) and DEVD-AMC (aceytl-Asp-Glu-Val-Asp-aminomethylcoumarin). Substrate hydrolysis is measured by determining the amount of fluorescent AMC product formation at excitation of 660 nm and emission at 460 nm with any fluorescent plate reader.[43] Specific inhibitors that will target selected apoptotic pathways are now being developed. It is hoped that these inhibitors will block individual Caspase proteases rather than a group of related enzymes. These inhibitors will allow more definitive identification of the protease(s) involved in the processing of specific substrates, as well as help in the ordering of upstream and downstream events within a particular apoptotic signaling pathway. Ultimately, these inhibitors, or related derivatives, may also result in more effective treatments of disease(s) associated with the aberrant regulation of apoptosis.

F.   A Hypothetical Model of Apoptotic Signaling Pathways

In Figure 7.3, a hypothetical working model of vertebrate apoptotic/PCD signaling is shown. This model incorporates much of what is known about the events that

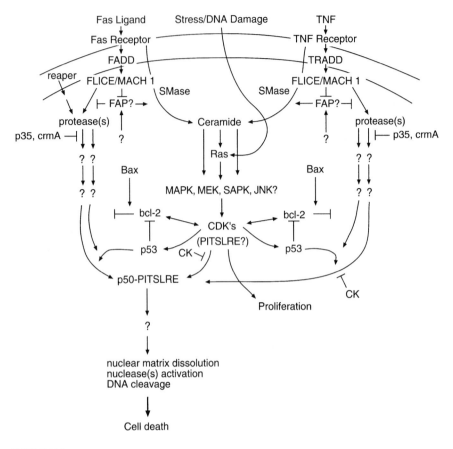

FIGURE 7.3

Hypothetical model of the signaling pathway(s) that may be involved in apoptosis: Multimerization of TNF and Fas receptors in response to their ligands results in the activation of Caspase protease cascade(s). Reaper (rpr) and stress are also capable of activating apoptotic protease pathway(s), possibly involving members of the MAP kinase family in response to ceramide production. The Fas receptor-associated phosphatase (FAP), as well as the p35 and crmA proteins, can block apoptotic signaling. In addition, members of the bcl-2 family can positively or negatively attenuate apoptotic signals as well.[16] Finally, members of the cyclin-dependent protein kinase (CDK) family may function as effectors downstream of the Caspase proteases.[56] The PITSLRE p110 protein kinase isoforms are substrates for the Caspase proteases, and the processed p50 PITSLRE isoform is an active histone H1 protein kinase.[57] Activation of this protein kinase is temporally associated with nuclear matrix dissolution and DNA cleavage, the so-called commitment phase of cell death from which there is no return.

accompany apoptosis, with particular emphasis on the receptor:protease pathway(s). Although some studies suggest that specific protein kinase(s) and phosphatase(s) may also be crucial components of apoptotic signaling pathway(s), the role of these proteins has not yet been confirmed or remain controversial.[56-59] Clearly, much more is known about the early apoptotic events associated with TNFR-related receptors (e.g., receptor interaction with TRADD, FADD/MORT1, and FLICE/MACHl) and the Caspase proteases, but the identification of their relevant target(s) should be forthcoming.[31] Proteins whose function is altered in a meaningful manner by the Caspase proteases are reasonable candidate downstream effector(s). As might be expected, apoptotic and programmed cell death pathways appear to be quite similar in some respects (e.g., Caspase enzyme activation by the *reaper* gene product), but there are also significant differences (e.g., reaper does not have an extracellular domain like the TNFR-related receptors, yet both activate Caspase enzyme activity). Even though protease cascades have been linked to reaper and the TNFR-related receptors, protease-independent signaling has been observed, suggesting that multiple, as well as possibly parallel, pathways may need to be triggered for cell death to occur.[60] The ability of the bcl-2 family of death antagonists/agonists to regulate certain death effectors (e.g., Bax signaling of specific cytoplasmic death components in the presence of Caspase inhibitors) independently of the Caspase enzymes supports this notion.[60] In addition, recent studies have demonstrated that multiple or even parallel protease cascades may be required to activate certain death effectors completely, committing cells to a point of no return.[61] The involvement of more than one effector pathway may have evolved to protect cells from inappropriately initiated death signals. The hierarchal control of the $p34^{cdc2}$ protein kinase, which prevents the inappropriate activation of this protein kinase if prerequisite cell cycle events (e.g., DNA replication) have not been completed, is an excellent example of such regulation.[62] As more is learned about the control of apoptosis, our view of the processes necessary for cellular suicide, as well as their relationship to other cellular processes, will evolve.

III. What Are Acceptable Endpoints for Determining Whether a Cell Is Undergoing Apoptosis/Programmed Cell Death?

A. DNA Fragmentation and Apoptosis

A common biochemical feature of apoptosis is the fragmentation of the condensed nuclear DNA.[63-66] In necrotic cells, chromatin condensation and margination do not occur, and these cells do not undergo autophagy; instead, scavenger cells (e.g., macrophages) are recruited to the site, where they proceed to engulf and digest the necrotic cellular remains. Early studies of apoptotic cells demonstrated that once they had committed to a death program, their nuclear DNA was cleaved by an activated endogenous nuclease, producing a "ladder" of DNA fragments of a defined periodicity.[63] In addition, digestion of chromosomal DNA during apoptosis may not be randomly targeted; instead, many of the linker regions targeted by apoptotic nucleases may be associated with the higher-order structure of chromatin.[65] Fragmented apoptotic DNA is very similar to the DNA fragmentation pattern observed when specific nucleases are used to cleave the spacer regions between nucleosomes.[67,68]

Limited nuclease digestion of chromatin generates oligonucleosomal DNA associated with histone octamers (i.e., nucleosomes), which can vary in length from 140 to 240 bp.[67,68] Depending on the amount of nuclease used to treat chromatin, as well as its incubation time, the resulting DNA fragments have a defined periodicity determined by the length of the DNA being protected by these histone octamers. The resulting nucleosome ladders are very similar to the oligonucleosomal ladders seen in many apoptotic cells. Cells undergoing PCD during normal development also share many of the cellular, molecular, and biochemical events associated with apoptosis, including the condensation and fragmentation of their DNA.[3,12] Such similarities suggest that many of the genes and/or proteins required for the execution of PCD during development may be identical to those required for apoptosis (e.g., the Caspase proteases and bcl-2).

In Figure 7.1 both CEM-C7 and Jurkat T cells generate oligonucleosomal DNA ladders during Fas-induced apoptosis (indicated by the +), but Molt-4 T cells do not. DNA ladder formation can be easily monitored by careful extraction of DNA and agarose gel electrophoresis.[63] The absence of DNA ladders in apoptotic MOLT-4 cells demonstrates that their formation does not always accompany cell death. In fact, Schulze-Osthoff and colleagues[6] have shown that enucleated murine fibrosarcoma cells (in which expression of the Fas receptor was directed by a stably integrated expression construct) undergo many of the morphological and biochemical changes normally associated with Fas receptor ligation. The authors of this study conclude that nuclear events (e.g., DNA fragmentation and nuclear signaling) are not necessarily required for apoptosis to occur. Furthermore, it has been shown that changes in cell membranes and mitochondria associated with apoptosis can occur in enucleated cytoplasts.[5,8,9] These conclusions may be valid for specific cell types, cell lines, and/or a specific set of apoptotic stimuli; however, there are many studies demonstrating that nuclear events are essential for apoptosis.[69,70] Thus, while changes in nuclear structure are associated with many different apoptotic programs, the extent and/or nature of DNA damage may reflect important differences in the effector(s) and/or apoptotic pathway(s) activated by different stimuli.

The development of new technologies has also had a significant impact on our ability to further define the events that are associated with apoptosis. For example, pulsed-field gel electrophoresis (PFGE) has been used to demonstrate that the DNA from apoptotic cells undergoes a much more restricted form of nuclease cleavage.[71] Limited nicking/cleavage of DNA can result in >50-kb and/or >300-kb DNA fragments that can be readily detected by PFGE. These larger DNA fragments seem to precede the appearance of the smaller oligonucleosomal DNA ladders during apoptosis. Furthermore, many apoptotic cells that do not form the oligonucleosomal DNA ladders (such as the human MOLT-4 T-cell line; Figure 7.1) have been shown to contain the larger 50- and/or 300-kb fragments. It has been suggested that this pattern of DNA cleavage may reflect specific differences in chromatin structure (e.g., sites of chromatin attachment to the nuclear matrix-lamina complex or nuclear scaffold) associated with various cell type.[72] It is also possible that distinct endonucleases (localized to specific cellular regions such as the nuclear matrix) or a single endonuclease that can be differentially activated by changes in the intracellular Ca^{2+} level (higher Ca^{2+} levels have been associated with DNA laddering) are responsible for differences in DNA fragmentation.

As an alternative to the analysis of oligonucleosomal DNA fragments by agarose gel electrophoresis, a method employing terminal deoxynucleotidyl transferase (TdT) was developed.[73] TdT-mediated dUTP-biotin nick end labeling (TUNEL) is now widely used for detecting and even quantitating either single- or double-strand nicks in cellular DNA. This technique is more versatile than agarose gel electrophoresis, since it allows *in situ* detection of programmed cell death at the single cell level, without disruption of the surrounding tissue. TUNEL can also be applied to cells in culture, allowing the detection of both intact (by BrdU labeling) and fragmented (by TUNEL) DNA by flow cytometry (FACS), providing a quantitative measure of apoptosis associated with a particular agent or signal. In addition, the simultaneous analysis of one or more additional proteins using monoclonal and/or polyclonal antibodies (which are either tagged by a fluorescent moiety or detected by the use of a fluorescently tagged second antibody) can be easily coupled to TUNEL assays.

B.   Caspase Protease Activities, the Nucleus, and Cell Death

One caveat regarding nuclease cleavage of chromosomal DNA during apoptosis is that cytoplasmic apoptosis can occur independently of nuclear events.[5,6,8] In these studies, the nucleus was removed from the cell prior to the administration of an apoptotic signal, yet the remaining cytoplasm underwent characteristic membrane blebbing. More recently, it has been shown that much of the cellular apoptotic machinery is contained within the cytoplasm, suggesting that nuclear components may not be necessary for the execution of cell death;[9,74] however, this point remains somewhat controversial, and it is not clear that the nucleus is excluded from all forms of apoptotic signaling. In fact, many of the known targets of ICE/CPP32-like proteases are nuclear proteins, including PARP, lamins, PKδ, topoisomerase, Ul-70K, and the retinoblastoma protein.[48,51,52,75,76] It is also possible that the nuclear changes associated with apoptosis are dependent upon cytoplasmic events occurring first, and that the transport of specific protein(s) into the nucleus is required for the execution of a nuclear death program. The vertebrate cell cycle is controlled in a similar manner, where translocation of cyclin regulatory subunits from the cytoplasm into the nucleus is required for the activation of the appropriate cdk catalytic subunit.[77] Finally, several groups have shown that apoptosis associated with the negative selection of autoreactive T cells is dependent on active transcription of genes such as Nur77.[69,70] Thus, even though cytoplasts can undergo apparent apoptosis, a nucleus may still be required for many cells to execute their cell death program properly.

The induction of ICE/CPP32-like (Caspase) protease activity has recently been shown to be associated with many forms of apoptosis.[13,78] Ten distinct Caspase proteases have been identified in mammals by molecular cloning, and their activities have been associated with the proteolytic processing of many different substrates.[13]

In many cases, the significance of the proteolytic processing of these substrates and apoptotic signaling has not yet been established. However, the processing of these substrates often coincides with morphological features indicative of apoptosis, such as DNA fragmentation, and they provide a convenient assay to measure cellular commitment to a death program. There is now evidence that cytoplasmic proteases (e.g., granzymes), can also be passively transported into the nucleus.[79] Once inside the nucleus, granzyme B appears to accumulate by binding to nuclear/nucleolar factors in a cytosolic factor-mediated manner. Localization of granzyme B to the nucleus was not accompanied by changes in either the active or passive nuclear transport properties of the cell, suggesting that the nuclear envelope and its constituents are not granzyme B substrates.

C.   Changes in the Localization of Membrane-Associated Phosphatidylserine During Apoptosis

Possibly one of the most reliable and consistent markers of apoptosis/PCD is the externalization of phosphatidylserine (PS) to the outer cell membrane.[8,80,81] Phosphatidylserine is a lipid normally localized to the inner side of the plasma membrane. However, PS is rapidly exported to the outer plasma membrane following induction of apoptosis by numerous stimuli, and it is thought that this relocalization of PS may provide a marker for recognition of apoptotic cells by macrophages. The relocalization of PS to the external plasma membrane during Fas-induced apoptosis will occur in the absence of a nucleus, suggesting that PS externalization precedes changes associated with the chromatin and nuclear matrix during apoptosis and that these changes are not required for this process.[8] Furthermore, apoptosis associated with PS externalization, in intact cells and cytoplasts, is associated with the induction of Caspase 3 enzyme activity and fodrin proteolysis, both of which can be blocked by the presence of Caspase 1/3 peptide inhibitors. Externalization of PS is easily detected by using fluorescently tagged annexin V, a Ca^{2+}-dependent, phospholipid-binding protein with high affinity for PS.[80] Use of fluorescently labeled (e.g., fluorescein isothiocyanate [FITC]) annexin V not only allows one to visualize apoptotic cells by IF, but it also allows the detection and quantitation of apoptotic cell death by fluorescence-activated cell sorter (FACS) analysis. Similarly, one can use different fluorescent tags (e.g., Hoechst 33258 or propidium iodide staining of cellular DNA, measurements of DNA fragmentation by use of a FITC-conjugated dUTP [TUNEL], 5-iodacetamidofluorescein labeling, and FITC or Texas Red-labeled annexin V) in conjunction with simultaneous analysis of other physiological parameters of apoptosis (e.g., cell size) to provide more definitive proof of cell death. Annexin V is available commercially as a biochemical reagent or as part of a kit designed for the detection of apoptotic cells. Thus far, there have been no reports of apoptosis in the absence of the externalization of PS. However, as with all of the previous parameters used to define apoptotic cells (such as DNA nicking and fragmentation), the possibility remains that an exception will be found for this marker as well.

IV. Analysis of Apoptosis Using Cell-Free Extract Systems

Identification of the proteins essential for the execution of programmed cell death and their interactions during this process in *C. elegans* and *Drosophila* has been facilitated by genetic analysis.[3,34] This approach is not readily feasible in vertebrates; therefore, analysis of programmed cell death in mammals has largely relied on the use of biochemical, molecular, and cellular analyses using established cell lines. The morphological changes accompanying apoptosis have been characterized in detail (see Section II); however, the biochemical mechanisms involved in the process are not as well understood due to the complex nature of apoptotic pathway(s).

Apoptosis proceeds in two phases: a commitment phase and an execution phase.[82,83] In the commitment phase, cells recognize a signal that commits them to a death program. Morphological features of apoptosis are not evident during the commitment, or "latent", phase. In addition, the latent phase is highly variable with respect to the nature of the apoptotic response(s) elicited by a variety of signals and their specific targets. The duration of this latent phase varies considerably, ranging from hours to days. Following the latent phase, cells enter what has been described as the "execution phase".[82] The events associated with the actual execution of cell death are much more rapid, and the morphological and biochemical changes associated with this phase appear to be very well conserved evolutionarily. The variable time frame of the commitment phase and the rapidity of the execution phase result in a general population of individual cells in various phases of apoptosis. Such cellular diversity complicates the biochemical and molecular analysis of specific protein participants, as well as the elucidation of the sequence of biochemical events that are taking place.

In vitro cell-free systems have been developed to study the biochemical pathways that constitute the execution phase of apoptosis.[82,84] Soluble cell extracts are prepared from activated cells already in the latent phase, partially circumventing the asynchrony problem encountered when using intact cells. In addition, the analyses are rapid and allow the determination of the role of specific components of apoptotic pathway(s) in isolation. The controlled removal of native endogenous components, or the addition of mutant forms of these same components, is also possible. Likewise, the effect of the addition of external elements (e.g., inhibitors or activators) is not hindered by the problem of membrane impermeability that is associated with some of these compounds when using intact cells. A remarkable feature of these cell-free systems is that the cytoplasmic extract is self-contained in its potential to trigger apoptosis, and it does not require the nucleus. This feature is consistent with earlier observations that apoptosis can take place in enucleated cells,[5,6] and permits the separate analysis of the effects of apoptosis on discrete compartments of the cell, vis-à-vis the nucleus and the cytoplasm.

Three types of cell-free extracts have been described from chicken, *Xenopus,* and mammalian cells,[84-87] all of which reproduce the execution phase of apoptosis *in vitro* (Figure 7.4). These systems use cytosolic extracts from activated cells to study the biochemical activities in the cytosol itself or to analyze the morphological and biochemical activities of isolated nuclei incubated in the presence of these cytosolic extracts (Figure 7.4) A hallmark event that occurs in the cytosol during apoptosis is the proteolytic activation of the Caspase enzymes, which can be detected by western blot analysis of these extracts. In the nucleus, apoptosis is characterized by biochemical changes that include DNA fragmentation, which, as mentioned earlier, can be detected by agarose gel electrophoresis as well as by a variety of more sophisticated experimental means. Morphological changes, such as chromatin condensation followed by membrane blebbing and the formation of apoptotic bodies, can be observed by microscopy.

A.   Chicken DU249 S/M Extracts

In the first *in vitro* system, extracts were prepared from chicken DU249 hepatoma cells,[85] after sequential synchronization in the S and M phases of the cell cycle (S/M extracts). The cell-free extracts are derived from essentially latent phase cells which are provoked into undergoing apoptosis by maintaining a lengthy drug-induced (aphidicolin) block at the Gl/S boundary. The S/M extracts maintain their apoptotic potential and show all of the characteristic features of apoptosis. For instance, nuclei incubated with these extracts exhibit DNA condensation, nuclear lamina breakdown, and DNA fragmentation, while control extracts derived from asynchronous cells do not. Some of the requirements for the morphological changes and DNA fragmentation have been determined by examining the effect(s) of the addition of exogenous components to these extracts, and they are noteworthy. Neither of the changes mentioned above require Ca^{2+}, ruling out the activation of a Ca^{2+}-dependent endonuclease. The presence of Mg^{2+} was shown to be essential for both of these *in vitro* systems, since the addition of EDTA abolished their apoptotic activities. Interestingly, the two activities can be separated by their sensitivity to Zn^{2+} and ATP. The morphological changes were prevented, but DNA fragmentation was unaffected in these experiments, suggesting that a Zn^{2+}-sensitive factor was responsible for the morphological changes only. The S/M extracts have also been used to demonstrate the processing of exogenous proteins that are biochemical components of the apoptotic pathway.[84] Poly(ADP-ribose) polymerase (PARP) was rapidly cleaved by the S/M extract, and this process was inhibited by the presence of YVAD-CHO, a peptide inhibitor of the ICE-related family of proteases.[88] Similarly, the role of the ICE-like protease, Mch2α (Caspase 6), in the cleavage of lamin A, was deduced from the inhibition profile using S/M cell extracts.[48]

B.   *Xenopus* Oocyte Cell-Free Extracts

The *Xenopus* cell-free extracts are made from frog eggs that are harvested 14 to 28 days after injecting frogs with hormone.[86] The physiological basis for the apoptotic activity of these egg extracts is undefined, but it may involve oocyte atresia that

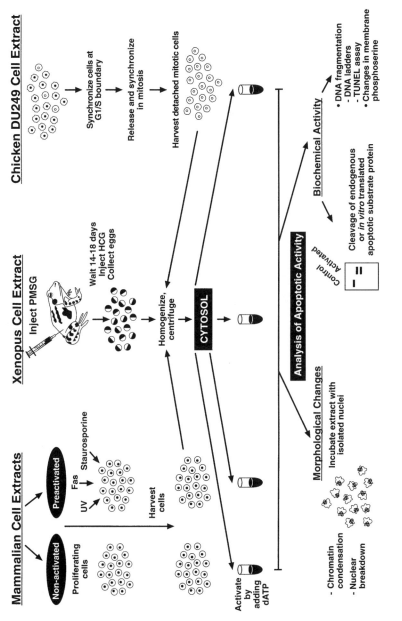

FIGURE 7.4

Schematic comparison of the various *in vitro* cell-free apoptotic assay systems: A step-by-step comparison of the components and requirements for the activation of each of the cell-free assay systems are shown (see text for details).

mimics the apoptotic process. Like the S/M extracts, the *Xenopus* extracts are also capable of morphological changes and DNA degradation using *Xenopus* sperm nuclei, but similar effects can also be seen when nuclei from different sources are used. Although both of these systems were insensitive to the presence of Ca^{2+}, the Ca^{2+} ionophore (ionomycin) blocked apoptosis, probably by releasing Ca^{2+} from their mitochondria or other cellular organelles. There are a few important differences that distinguish the *Xenopus* and S/M cell-free systems. A heavy membrane fraction enriched in mitochondria is essential for apoptotic activity in the *Xenopus* system, whereas the soluble fraction was sufficient in the S/M system. In addition, the *Xenopus* system was responsive to inhibition by baculovirus-synthesized bcl-2 protein, a potent inhibitor of apoptosis, while similar inhibition by bcl-2 was not documented in the S/M extracts.

C.   Mammalian Cell-Free Extracts

The third cell-free extract system uses mammalian cell lines as a source of the cytosolic extracts. Many different cell types, as well as protocols, have been described for preparing these extracts.[84,87,88] Based on the method of preparation, mammalian cell-free extracts are broadly defined as being one of two types. In the first system, extracts can be derived from various cell lines which can then be exposed to a variety of apoptotic stimuli prior to processing. These cells are then harvested, following a brief pre-incubation period to allow any apoptotic priming processes to take place, but before any characteristic apoptotic features are exhibited. A variety of agents that initiate apoptosis, including ultraviolet irradiation (UV), Fas antibody ligation, the protein tyrosine kinase inhibitor staurosporine, and the DNA topoisomerase I inhibitor campothecin (which represent only a few of a growing list of such agents), have been successfully employed to obtain cell-free apoptotic extracts.[87] These extracts reproduce the morphological and biochemical features that normally accompanying apoptosis. Fas- or UV-induced extracts from Jurkat cells were protected from apoptosis by ectopically expressed bcl-2 protein, as well as the peptide inhibitors of the Caspase proteases.[87]

The cell-free extracts also provide a means of testing whether the initiation of apoptosis by agents such as Fas mAb or UV-irradiation can be bypassed and alternatively triggered by the addition of "intermediate" components of these apoptotic pathways. Induction of ceramide, a lipid second messenger, has been associated with a number of apoptotic signaling pathways that occur in many different cell types.[87] Incubation of mammalian cell extracts, prepared from uninitiated cells, with ceramide can elicit a significant apoptotic response. This suggests that such "intermediate" compounds are capable of replacing or mimicking signals generated by primary initiators (e.g., Fas mAb or UV irradiation). Similar results were observed in the *Xenopus* cell-free system following depletion of the essential heavy membrane fraction (an important source of these components), demonstrating that exogenously added ceramide was capable of restoring the apoptotic potential of the depleted extracts.

The second type of mammalian cell-free extract system also utilizes extracts derived from logarithmically growing, untreated cells.[88] In this case, however, initiation of the apoptotic program was triggered by the addition of deoxyadenosine triphosphate (dATP) to the extract. Nuclei underwent morphological changes, DNA was fragmented, and the Caspase proteolytic cleavage of various substrates proceeded in a manner similar to the other cell-free extract systems described above. Fractionation of the dATP-activated extracts was used to identify cytochrome C as one of the essential factors involved in the induction of apoptosis in this particular system. Since cytochrome C is a mitochondrial protein, a common theme appears to be emerging which underscores the role of mitochondrial components in most of the cell-free systems.

How can the *in vitro* cell-free systems be used to study apoptosis? As mentioned earlier, apoptosis can be initiated by a variety of agents, and its execution may involve pathways of signal transduction unique to the inducer, while other components of these pathway(s) may be shared by several inducers. Cell-free extracts are particularly useful in determining whether a particular component, activated in response to a specific apoptotic trigger, has a unique role in apoptosis or whether this component has a more general role (i.e., a common factor) in apoptotic pathways induced by various triggers. For example, proteolytic cleavage of proteins such as fodrin, PARP, and some of the Caspase proteases are events that occur in cells that have been exposed to Fas mAb, TNFα, UV or γ-irradiation, and ceramide.[83,84,89] The ability to add components of a particular apoptotic pathway back into the cell-free extracts can be exploited to determine the position of a specific component in the hierarchy of apoptotic signal transduction. Cell-free extracts from normal cells have been used to demonstrate that granzyme B, by itself, can initiate all of the cellular apoptotic events through its cleavage of Caspase 3.[30,90] This observation has placed granzyme B upstream of the Caspase 3 protease. Furthermore, the dissection of individual biochemical steps is also amenable to analysis using inhibitors of the Caspase proteases. As an example, the cleavage of Caspase 3 by granzyme B has been shown to involve a two-step process involving two distinct Caspase proteases.[30] The first step was inhibited by the viral protease inhibitor, crmA, while the second step was specifically inhibited by DEVD-CHO, a Caspase 3-specific peptide inhibitor. Addition of only one of these inhibitors to apoptotic cell extracts selectively inhibited a subset of the cellular and molecular events associated with cell death, but not all. Conversely, addition of the second inhibitor, by itself, affected a different subset of events. As expected, the addition of both inhibitors broadly affected the events associated with apoptosis. These studies support the hypothesis that parallel Caspase/effector signaling pathways may exist, and may even be required, for the proper execution of apoptosis in some cell types (see Figure 7.3) Activated cell-free extracts have also been used to identify and/or confirm the existence of novel Caspase proteases. Five different proteins have been identified thus far in S/M extracts using the YV(bio)KD-amk peptide, which labels the active site of several of the Caspase proteases.[84]

Cell-free systems are extremely useful tools that can be used to identify and functionally evaluate novel effectors of apoptotic signaling. A cautious approach is essential in extrapolating the conclusions obtained from *in vitro* systems to the actual processes *in vivo*. Cell-free systems deal with soluble components of the cell as well as intracellular interactions; however, the role of the insoluble and ultrastructural components of a cell and their interaction(s) with soluble components are not addressed. These interactions may be important in the regulation of apoptosis and determining the rate of the response. Cell-free extracts also suffer from difficulties encountered in attempts to determine whether the events that occur in one system are equivalent to the events that occur in another system. Among the present systems available, the chicken cell extract mimics cells undergoing post-mitotic apoptosis, while the *Xenopus* system mimics oocyte atresia. Mammalian extracts can be prepared from cells triggered with a wide variety of apoptotic signals, including physiological signals such as Fas receptor ligation or glucocorticoid receptor activation, external agents such as UV or γ-irradiation, or chemical inhibitors of cellular proteins such as staurosporine, taxol, and campothecin. Although the basic mechanism of apoptosis is thought to be largely conserved, species- and/or trigger-specific differences may exist. The manner of extract preparation, as well as the composition of the isolation buffers used, may also contribute, at least in part, to the differences observed between these *in vitro* systems.

Despite these concerns, cell-free systems offer unique advantages, and they can be very helpful tools for the analysis of the myriad of processes involved in apoptotic signaling. New information regarding the apoptotic process obtained from these cell-free extract systems can then be used to design relevant *in vivo* experiments or to complement or confirm existing information obtained from *in vivo* cell line and animal models.

V.   Final Comments

Clearly, the events that mediate apoptosis and programmed cell death are complex and in many ways resemble mitogenic pathways. One of the most unique features of these pathways, thus far, is the involvement of a large family of related cysteine proteases.[13,78] Unlike the cell surface receptors commonly associated with mitogenic signaling pathways, the receptors that trigger death do not contain cytoplasmic protein kinase domains, and they do not associate directly with other cytoplasmic protein kinases. Instead, apoptotic signals appear to be attenuated through protease cascades and alterations in the composition of mitochondrial and nuclear membrane pores. Downstream targets of these proteases, including protein kinases, the DNA repair enzyme poly(ADP-ribose) polymerase, the cytoskeletal protein α-fodrin, the nuclear lamins, the 70-kDa protein component of the U1 small nuclear ribonucleoprotein, and the retinoblastoma protein, are either potential effectors of apoptotic signaling or proteins whose inactivation might be required for the demise of a cell. Proteolytic processing of these targets most likely involves multiple proteases, as well as phosphorylation and/or dephosphorylation. To complicate matters further, protease- and nuclear-independent apoptotic deaths have been observed.[5,6,8,9,60] Many would now suggest that *de novo* expression of genes or the nucleus itself is not necessary for apoptosis.[10] Determining whether active nuclear events are required for the proper execution of apoptosis is an issue that has not been resolved. Furthermore, is cell death universal, and what are the basic requirements for its execution? Do apoptotic and mitogenic signaling pathways overlap, and do they share common elements? Are apoptotic signaling pathways circumvented in tumor cells? It is certainly possible that death can come in many disguises, but as we remove these masks the faces may be familiar.

Acknowledgments

Regretfully, due to the limitations of space a number of citations for published work could not be included, and we apologize to these investigators. The authors would like to acknowledge the support of the NIH (GM44088 and CA67938 to VJK and CA21765 to St. Jude Children's Research Hospital), and the support of the American Lebanese Syrian Associated Charities (ALSAC) to VJK and JML for research done in their laboratories.

References

1. Kerr, J. F. R., Wylie, A. H., and Currie, A. R., *Br. J. Cancer,* 26, 239, 1972.
2. Wylie, A. H., Kerr, J. F. R., and Currie, A. R., *Int. Rev. Cytol.,* 68, 251, 1981.
3. Ellis, R. E., Yuan, J., and Horvitz, H. R., *Ann. Rev. Cell Biol.,* 7, 663, 1991.
4. Williams, G. T. and Smith, C. A., *Cell,* 74, 777, 1993.
5. Jacobson, M. D., Burne, J. F., and Raff, M. C., *EMBO J.,* 13, 1899, 1994.
6. Schulze-Osthoff, K., Walczak, H., Droge, W., and Krammer, P. H., *J. Cell Biol.,* 127, 15, 1994.
7. Enari, M., Hase, A., and Nagata, S., *EMBO J.,* 14, 5201, 1995.
8. Martin, S. J., Finucane, D. M., Amarante-Mendes, G. P., O'Brien, G. A., and Green, D. R., *J. Biol. Chem.,* 271, 28753, 1996.
9. Weil, M., Jacobsen, M. D., Coles, H. S. R. et al., *J. Cell Biol.,* 133, 1053, 1996.
10. Raff, M. C., Barres, B. A., Burne, J. F., Coles, H. S., Ishizaki, Y., and Jacobson, M. D., *Science,* 262, 695, 1993.
11. Fraser, A. and Evans, G., *Cells,* 85, 781, 1996.
12. Suston, J. E. and Horvitz, H. R., *Dev. Biol.,* 82, 110, 1977.
13. Yuan, J., *Curr. Opin. Cell Biol.,* 7, 211, 1995.
14. Metzstein, M. M., Hengartner, M. O., Tsung, N., Ellis, R. E., and Horvitz, H. R., *Nature,* 382, 545, 1996.
15. Ellis, R. E., Jacobson, D. M., and Horvitz, H. R., *Genetics,* 129, 79, 1991.

16. Oltvai, Z. N. and Korsmeyer, S. J., *Cell,* 79, 189, 1994.

17. Sherr, C. J. and Roberts, J. M., *Genes Dev.,* 9, 1149, 1995.

18. Sherr, C. J., *Cell,* 79, 551, 1994.

19. Lydall, D. and Weinert, T., *Science,* 270, 1488, 1995.

20. Yonish-Rouach, E., Grunwald, D., Wilder, S. et al., *Mol. Cell. Biol.,* 13, 1415, 1993.

21. Brugarolas, J., Chandrasekaran, C., Gordon, J. I., Beach, D., Jacks, T., and Hannon, G. J., *Nature,* 377, 552, 1995.

22. Raff, M. C., *Nature,* 356, 397, 1992.

23. Ashwell, J. D., Berger, N. A., Cidlowski, J. A., Lane, D. P., and Korsmeyer, S. J., *Curr. Biol.,* 15,147, 1994.

24. Zeitlin, S., Lin, J.-P., Chapman, D. L., Papaioannou, V. E., and Efstratiadis, A., *Nature Genet.,* 11, 155, 1995.

25. Green, D. R. and Martin, S. J., *Cur. Opin. Immunol.,* 7, 694, 1995.

26. Yuan, J., Shaham, S., Ledoux, S., Ellis, H. M., and Horvitz, H. R., *Cell,* 75, 641, 1993.

27. McConkey, D. J., Jondal, M., and Orrenius, S., *Semin. Immunol.,* 4, 371, 1992.

28. Inoue, S., Sharma, R. C., Schimke, R. T., and Simoni, R.D., *J. Biol. Chem.,* 268, 5894, 1993.

29. Alnemri, E. S., Livingston, D. J., Nicholson, D. W. et al., *Cell,* 87, 171, 1996.

30. Martin, S. J., Amarante-Mendes, G. P., Shi, L. et al., *EMBO J.,* 15, 2407, 1996.

31. Cleveland, J. L. and Ihle, J. N., *Cell,* 81, 479, 1995.

32. Zhang, J. and Winoto, A., *Mol. Cell. Biol.,* 16, 2756, 1996.

33. Hsu, H., Xiong, J., and Goeddel, D. V., *Cell,* 81, 495, 1995.

34. White, K., Grether, M. E., Abrams, J. M., Young, L., Farrell, K., and Stellar, H., *Science,* 264, 677, 1994.

35. Chen, P., Lee, P., Otto, L., and Abrams, J., *J. Biol. Chem.,* 271, 25735, 1996.

36. Hay, B. A., Wassarman, D. A., and Rubin, G. M., *Cell,* 83, 1253, 1995.

37. Li, P., Allen, H., Banerjee, S. et al., *Cell,* 80, 401, 1995.

38. Kuida, K., Lippke, J. A., Ku, G. et al., *Science,* 267, 2000, 1995.

39. Kuida, K., Zheng, T. S., Na, S. et al., *Nature,* 384, 368, 1996.

40. Bakhshi, A., Jensen, J. P., Goldman, P. et al., *Cell,* 41, 899, 1985.

41. Cleary, M. L. and Sklar, J., *Proc. Natl. Acad.Sci., USA,* 82, 7439, 1985.

42. Tsujimoto, Y., Gorham, J., Cossman, J., Jaffe, E., and Croce, C. M., *Science,* 229, 1390, 1985.

43. Armstrong, R. C., Aja, T., Xiang, J. et al., *J. Biol. Chem.,* 271, 16850, 1996.

44. Garcia, I., Martinou, I., Tsujimoto, Y., and Martinou, J. C., *Science,* 258, 302, 1992.

45. Bissonnette, R. P., Echeverri, F., Mahboubi, A., and Green, D. R., *Nature,* 359, 552, 1992.

46. Vaux, D. L., Weissman, I. L., and Kim, S. K., *Science,* 258, 1955, 1992.

47. Muchmore, S. W., Sattler, M., Liang, H. et al., *Nature,* 381, 335, 1996.

48. Lazbnik, Y. A., Kaufmann, S. H., Desnoyers, S., Poirier, G. G., and Earnshaw, W. C., *Nature,* 371, 347, 1994.

49. Xue, D. and Horvitz, H. R., *Nature,* 377, 248, 1995.

50. Ray, C. A., Black, R. A., Kronheim, S. R. et al., *Cell,* 69, 597, 1992.

51. Tewari, M., Beidler, D. R., and Dixit, V. M., *J. Biol. Chem.,* 270, 18738, 1995.

52. Emoto, Y., Manome, Y., Meinhardt, G. et al., *EMBO J.,* 14, 6148, 1995.

53. Liston, P., Roy, N., Tamai, K. et al., *Nature,* 379, 349, 1996.

54. Boudreau, N., Sympson, C. J., Werb, A., and Bissell, M. J., *Science,* 267, 891, 1995.

55. Milligan, C. E., Prevette, D., Yaginuma, H. et al., *Neuron,* 15, 385, 1995.

56. Shi, L., Nishioka, W. K., Th'ng, J., Bradbury, E. M., Litchfield, D. W., and Greensburg, A. H., *Science,* 263, 1143, 1994.

57. Lahti, J. M., Xiang, J., Health, L. S., Campana, D., and Kidd, V. J., *Mol. Cell Biol.,* 15, 1, 1995.

58. Norbury, C., MacFarlane, M., Fearnhead, H., and Cohen, G. M., *Biochem. Biophys. Res. Comm.,* 202, 1400, 1994.

59. Zanke, B. W., Boudreau, K., Rubie, E. et al., *Curr. Biol.,* 6, 606, 1996.

60. Xiang, J., Chao, D. T., and Korsmeyer, S. J., *Proc. Natl. Acad. Sci. USA,* 93, 14559, 1996.

61. Lazebnik, Y. A, Takahashi, A., Moir, R. D. et al., *Proc. Natl. Acad. Sci. USA,* 92, 9042, 1995.

62. Hartwell, L., *Cell,* 71, 543, 1992.

63. Williams, J. R., Little, J. B., and Shipley, W. U., *Nature,* 252, 754, 1974.

64. Compton, M., *Cancer Metastasis Rev.,* 11, 105, 1992.

65. Filipski, J., Leblanc, J., Youdale, T., Sikorsksa, M., and Walker, P. R., *EMBO J.,* 9, 1319, 1990.

66. Bortner, C. D., Oldenburg, N. B. E., and Cidlowski, J. S., *Trends in Cell. Biol.,* 5, 21, 1995.

67. Prunell, A., Kornberg, R. D., Lutter, L., Klug, A., Levitt, M., and Crick, F. H. C., *Science,* 204, 855, 1979.

68. Butt, T. R., Jump, D. B., and Smulson, M. E., *Proc. Natl. Acad. Sci. USA,* 76, 1628, 1979.

69. Woronicz, J. D., Calnan, B., Ngo, V., and Winoto, A., *Nature,* 367, 281, 1994.

70. Liu, Z.-G., Smith, S. W., McLaughlin, K. A., Schwartz, L. M., and Osborne, B. A., *Nature,* 367, 281, 1994.

71. Oberhammer, F., Wilson, J. W., Dive, C. et al., *EMBO J.,* 12, 3679, 1993.

72. Walker, P. R., Pandey, S., and Sikorska, M., *Cell Death Diff.,* 2, 97, 1995.

73. Gavrieli, Y., Sherman, Y., and Ben-Sasson, S. A., *J.Cell Biol.,* 119, 493, 1992.

74. Jacobsen, M. D., Weil, M., and Raff, M. C., *J. Cell Biol.,* 133, 1041, 1996.

75. Takahashi, A., Alnemri, E. S., Lazbnik, Y. A. et al., *Proc. Natl. Acad. Sci. USA,* 93, 8395, 1996.

76. Janicke, R. U., Walker, P. A., Lin, X. Y., and Porter, A. G., *EMBO J.,* 15, 6969, 1996.

77. Pines, J. and Hunter, T., *J. Cell Biol.,* 115, 1, 1991.

78. Henkart, P. A., *Immunity,* 4, 195, 1996.

79. Jans, D. A., Jans, P., Briggs, L. J., Sutton, V., and Trapan, J. A., *J. Biol. Chem.,* 271, 30781, 1996.

80. Vermes, I., Haanen, C., Steffens-Nakken, H., and Reutelingsperger, C., *J. Immunol. Methods,* 184, 39, 1996.

81. Vanags, D. M., Porn-Ares, M. I., Coppola, S., Burgess, D. H., and Orrenius, S., *J. Biol. Chem.,* 271, 31075, 1996.

82. Clarke, A. R., Sphyris, N., and Harrison, D. J., *Mol. Med. Today,* 1, 189, 1996.

83. Earnshaw, W. C., *Curr. Opin. Cell Biol.,* 7, 337, 1995.

84. Earnshaw, W. C., *Trends Cell Biol.,* 5, 217, 1995.

85. Lazebnik, Y. A., Cole, S., Cooke, C. A., Nelson, W.G., and Earnshaw, W. C., *J. Cell Biol.,* 123, 7, 1993.

86. Newmeyer, D. D., Farschon, D. M., and Reed, J. C., *Cell,* 79, 353, 1994.

87. Martin, S. J., Newmeyer, D. D., Mathias, S. et al., *EMBO J.,* 14, 5191, 1995.

88. Liu, X., Kim, C.N., Yang, J., Jemmerson, R., and Wang, X., *Cell,* 86, 147, 1996.

89. Hannun, Y. A., *Science,* 274, 1855, 1996.

90. Darmon, A. J., Nicholson, D. W., and Bleackley, R. C., *Nature,* 377, 446, 1995.

Chapter 8

An *In Vitro* Model for Human Bladder Cancer Pathogenesis Studies

Thomas R. Yeager, David F. Jarrard,
and Catherine A. Reznikoff

Contents

0-8493-4411-5/98/$0.00+$.50

I.   Introduction

Bladder cancer is currently the fifth most common type of cancer in the U.S. The American Cancer Society estimates that 53,000 new cases and 12,000 deaths will be attributed to bladder cancer in 1996. These numbers represent a significant health problem. Overall, bladder cancer affects men three times more often than women, and its incidence increases with age.

    Bladder cancer has been intensively studied during the past century as a model for chemical carcinogenesis.[1] In the late 1800s, workers in the rubber and dye industries were found to have a very high incidence of bladder cancer. Later studies showed that individuals with occupational exposure to certain classes of chemicals had a bladder cancer incidence up to 50 times higher than nonexposed individuals. Arylamines, including 4-aminobiphenyl and 2-napthylamine, were specifically identified in the 1950s as the etiologic agents responsible for the increased bladder cancer incidence observed in certain industrial settings. More recently, cigarette smoking has been correlated with doubling the risk for bladder cancer. The increased risk for bladder cancer attributed to smoking may result from exposure to some of the same chemicals identified in cigarette smoke (e.g., arylamines) that are associated with bladder cancer etiology in industrial settings. Notably, the anatomy of the bladder facilitates the bladder's exposure to carcinogens, as these are concentrated in urine which bathes the bladder epithelium before voiding.

    Other factors, such as bacterial infection, have also been associated with bladder cancer etiology.[2] In Egypt, infection with the parasite *Schistosoma hematobium* is thought to be responsible for a large number of bladder cancer cases. DNA from high risk human papillomavirus (HPV), usually HPV16, has been detected in about one third of bladder cancers, and its presences correlates with a high tumor grade.

    In North America approximately 90% of bladder tumors are transitional cell carcinoma (TCC), with 5% being squamous cell carcinoma (SCC) and 1% adenocarcinoma.[1] In Egypt, where schistosomiasis infection is very common, SCC makes up approximately 50% of all bladder cancers. The overwhelming majority of TCCs at diagnosis are papillary and of low grade. These frequently recur, but rarely progress (≈20%). To date, there are no definitive markers to distinguish progression. The 5-year survival rates for the low-grade papillary TCC is over 90%. In contrast, the

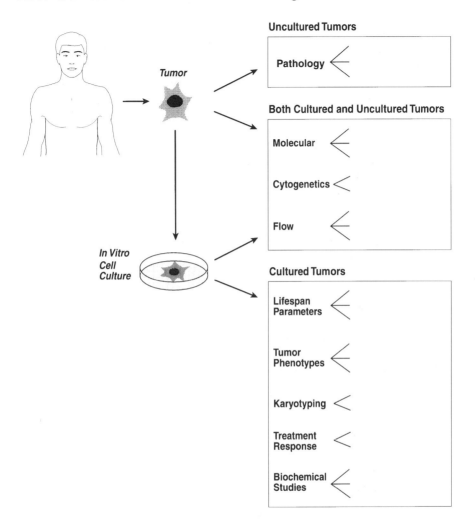

FIGURE 8.1
Shown here are examples of assays that can be done with uncultured and/or cultured tumor samples.

5-year survival rates for invasive TCC with regional or distant disease spread are 49% and 6%, respectively.

Recent reviews have documented much genetic information gained from direct analysis of bladder cancer biopsies (Figure 8.1).[2,3] Restriction fragment length polymorphism (RFLP) analysis has been particularly useful in identification of genetic losses. More recently, comparative genomic hybridization (CGH) has been used to identify both genetic gains and losses. The most frequent alteration observed to date is the chromosome 9 loss, which has been associated with early stage TCC. Additional frequent chromosomal losses are –3p, –8p, –11, –12, –13q, and –17p, while significant gains include +1, +3q, +5p, +7, +8q, and +13. These cytogenetic losses and gains are thought to signify either the deletion/inactivation of tumor supressor genes (TSGs), such as *CDKN2/p16* at 9p21, *RB* at 13q14, or *TP53* at 17p13, or amplification of oncogenes such as epidermal growth factor receptor (*EGFR*) on 7p or ERBB-2 on 17q. Other losses and gains may signify locations of as yet undiscovered genes that could play critical roles in cancer pathogenesis. In summary, analysis of uncultured tumor material has led to identification of regions in which genetic alterations occur during tumorigenesis, but there are limitations to this approach.

A complementary approach to direct analysis of tumor material would be to study the molecular and cellular phenotypes of the tumors using *in vitro* cell culture methods (Figure 8.1). These studies might include assessment of the ability of the tumor cells to form immortal cell lines, determination of tumor cell growth in soft agar, tumor growth assessment in athymic nude mice, analysis of chromosome alterations by karyotyping and fluorescent *in situ* hybridization (FISH), and tumor cell response to a variety of therapeutic regimens including irradiation and chemotherapy. Thus, the ability to culture tumor cells *in vitro* not only expands the sample size and allows for molecular evaluation, but also permits the assessment of complex phenotypic parameters that are not possible with primary samples.

Cell culture models permit unique clinically relevant questions to be addressed.[1] For example, cell culture allows unique approaches to addressing difficult questions concerning the prognosis of human bladder tumors. For example there are no genetic or histopathologic markers to distinguish which papillary TCCs will progress and become invasive from those that will remain superficial and benign. If an *in vitro* marker for progression were to be identified, this would be very useful clinically. In addition, *in vitro* assays could provide useful information to direct the urologic oncologist in the selection of classes of drugs for chemotherapy or to determine if irradiation is a rational choice for treatment.

II. Establishment of Transitional Cell Carcinoma Cell Cultures

A.   Sample Procurement

Transitional cell carcinoma samples are procured from the operating room utilizing several different approaches. Typically, a transurethral bladder tumor resection is performed using a cystoscope and a wire loop containing electrocautery. Papillary lesions are debulked using this loop in the absence of electricity ("cold loop"), and the tissue is obtained by using cotton or wire mesh to strain when the bladder is emptied. The base of the bladder and adjoining muscle are resected separately and sent for permanent pathological analysis. Sessile lesions and carcinoma *in situ* (CIS) may often be obtained utilizing a cup biopsy, with additional resection of the lesion done using the cautery loop. Avoiding the use of electrocautery during sample procurement improves the culture viability of the specimen. An additional source of tissue is obtained at the time of cystectomy, or removal of the bladder for cancer.

After the specimen is removed, the bladder is opened in a sterile fashion and obvious tumor removed using a scalpel. One critical aspect of this specimen procurement is providing adequate tissue for permanent pathologic analysis. In general, tumors less than 1 cm, unless there are multiple lesions, are not harvested for culture and are sent solely for histology.

Tissue to be sent for culture is placed in sterile, refrigerated Ham's F12 medium supplemented with 1% fetal bovine serum (FBS) and growth factors (1%FBS-F12+) and taken to the laboratory.[4] If adequate tissue is available, a portion of this is flash frozen in liquid nitrogen for correlative molecular studies (see Figure 8.1) This is also available should additional tissue be required for diagnostic purposes for the pathologist. Tissues sent for permanent pathologic analysis are generally adjacent to the samples for culture and provide the grade and stage of the tumors. Staining of frozen tissue by hematoxylin and eosin (H&E) after cryostatic sectioning provides a complementary source of histologic information.[1] An additional source of control tissue for molecular studies is also available from peripheral blood by collecting normal blood lymphocytes.

Tissue sources for normal human uroepithelial cells (HUC) have been obtained primarily from ureteral specimens.[4-7] During renal harvest and transplantation, the distal ureter is often not utilized and is therefore transected and discarded. This tissue is sent to the laboratory after being placed in Ham's 1%FBS-Fl2+ medium. The donors of kidneys for transplantation have no history of bladder cancer, thus the likelihood of these specimens containing tumor tissue is small.

B. Cell Culture Methods

Tumor samples are set up for tissue culture as soon as possible to minimize loss of viability. Samples may be stored at 4°C, but the best results have been achieved when the tumor is processed without delay. The initial culture of TCC samples is performed in a fashion similar to HUC culture. To prepare a tumor for tissue culture, the sample is cut into 1-mm^2 pieces in a glass Petri dish containing a small volume of 1%FBS-Fl2+ using a sterile scalpel while the tumor is held with sterile forceps.[4] After the tumor is cut into explants, the explants are placed onto collagen-coated culture dishes, ideally using 10 explants for a 60-mm dish and 30 explants for a 100-mm dish. Minimal amounts of growth medium are used when the explants are initially placed on the dishes (e.g., 1 to 2 ml of medium per l00-mm dish). Using a small volume of medium allows the explants to attach better because they do not float. In the next several days, medium is gradually added to bring the final volumes to 10 ml per 100-mm dish and 5 ml per 60-mm dish. Outgrowth from the explants can be detected as early as 24 hr after the initial plating, depending on the viability of the tumor. Papillary tumors usually show faster initial outgrowth than invasive tumors. The growth medium is subsequently routinely renewed three times a week. Several factors influence the extent of outgrowth, including the method of tumor procurement (samples collected by cauterization typically show less growth than samples collected using the cold loop method) and the time until explants are cultured.

To disperse cells for serial passage in culture, cryopreservation, or experimentation (Figure 8.1), a nonenzymatic technique is employed.[4] Cultured TCCs usually require passage after 7 to 10 days when they reach 60 to 80% confluence. Briefly, for passage, cells are rinsed with 5 ml per 100-mm dish of 0.1% EDTA in Mg^{2+}- and Ca^{2+}-free Hanks balanced salt solution (HBSS). Next, the EDTA is removed and replaced with the same volume of fresh EDTA, and the dishes are placed in a 37°C rotary shaker for 10 to 15 minutes. Then, a pipette is used to rinse and gently loosen cells from the collagen substrate. The cells are collected in the EDTA solution and finally the plates are rinsed with an equal amount of growth medium to collect any residual cells. All cells collected from each culture dish are pooled in a centrifuge tube containing a final volume of 10% FBS and spun at 900 rpm (200 g) for 5 min. The supernate is removed, and the pelleted cells are then resuspended in complete medium using gentle pipetting to break up any clumped cells. Samples of the single cell suspension can be stained with 0.1% trypan blue to count viable cells using a hemocytometer for quantitative experimentation. For cryopreservation, 10% dimethylsulfoxide (DMSO) and 10% FBS are added to the growth medium and cells are distributed at 1×10^6 cells per freezing vial. Best results are obtained when cells are slowly cooled (2-hr freezing time) to the final freezing temperature of liquid nitrogen.

C.   Assessment of Tumor Cell Growth Potential *In Vitro*

The potential for extended lifespan or immortalization can be assessed after the normal growth period of 4 to 7 passages (P), using a 1–2 or 1–3 split, is reached.[8-10] Typically TCC cultures proliferate for 2 to 5 passages and then enter a crisis phase during which cells senesce. At crisis, cells should be passed from a 100-mm dish into smaller dishes (60- or 30-mm) for the purpose of maintaining sufficient cell density to support growth. A useful technique for success is to leave as many tumor explants in the culture as possible when passed. These explants will continue to be the source of new cell growth. Once most cultures have reached the crisis period, the cell numbers may remain unchanged for many weeks. During this period, all cells may senesce; alternatively, some or all cells may continue to proliferate and give rise to a potentially mortal cell line. The latter possibility suggests the presence of an immortal population *in vivo*.

D.   Use of Viral Oncogenes for Immortalization

In cases where TCCs do not spontaneously form cell lines, but an immortal line is desired for experimentation purposes, immortalization can be accomplished using viral oncogenes. In our lab, this has been done using the HPV16 E6 and/or E7 viral oncogenes.[8,9] Since about one third of bladder tumors have been shown to contain HPV16, this is a relevant virus to use in immortalization of this cell type. The protein products of the E6 and E7 genes bind to and inactivate the p53 and pRb tumor suppressor gene products (TSGs), respectively. Use of the E6 and E7 genes substitutes for mutation of these important TSGs, which are among the most frequently mutated genes in bladder cancers. HPV16 E6 and/or E7 routinely extend the lifespan of HUC and TCC, but immortalization occurs at a low frequency.

The HPV oncogenes can most efficiently be delivered into cells using retroviral vectors (constructed by Denise Galloway at Fred Hutchinson Cancer Research Center).[8,9] Cells are infected in log phase growth using a virus/cell multiplicity of 0.1. TCC infected with these viruses may form immortal lines, when uninfected control TCC spontaneously senesce. Several laboratories have also used HPV18 for immortalization of human epithelial cells. In addition to using HPVs, the simian virus 40 (SV40) has been used extensively for immortalization of many different human cell types.[11,12] The mechanism of immortalization is similar to HPV, as the SV40 large T-antigen oncoprotein also binds and inactivates the p53 and pRb TSG products.

III. *In Vitro* Assays for Cancer Pathogenesis Studies

A. Biological Endpoints of Cellular Transformation

In vitro biological endpoints can be useful parameters to judge tumor aggressiveness *in vivo*. Such parameters might include cell morphology, lifespan potential, ability to form colonies in a semi-solid substrate, and tumorigenicity in athymic nude mice. These parameters may complement tumor pathology and correlate with or predict tumor progression.

1. Morphological alterations

An obvious characteristic of a TCC culture is its ability to recapitulate normal HUC morphology.[5,8,10] Normal HUCs grow out from the explant and form a tight continuous sheet of cuboidal cells that stratify and form two to three cell layers at confluence (Figures 8.2A and B). Low grade TCCs sometimes show a morphology similar to normal HUC. In contrast, other TCCs can exhibit significantly altered morphology, forming multilayered structures *in vitro* reminiscent of the papillary growths they formed *in vivo* (Figures 8.2C and D). The potentially more invasive CIS bladder cells often form monolayers that are loosely adherent and show altered cellular morphology compared to HUCs (Figures 8.2E and F).

2. Spontaneous immortalization *in vitro*

The lifespan of TCC *in vitro* using the conditions described above varies from two to seven passages, compared to HUCs that routinely show three passages representing

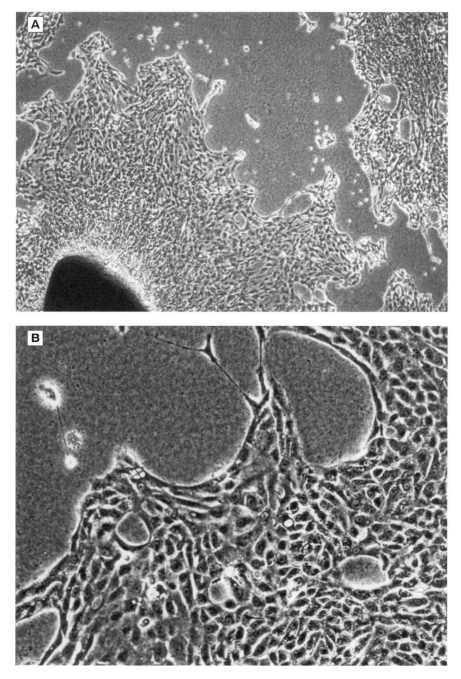

FIGURE 8.2
Comparative morphology of HUC and TCC samples in culture: HUC cultures show a tight epithelial morphology, 40× **(A)** and 200× **(B)**. Papillary TCC cultures also show a tightly packed epithelial morphology, 40× **(C)** and 200× **(D)**, that is similar to HUCs. Note that the cellular outgrowth from explants

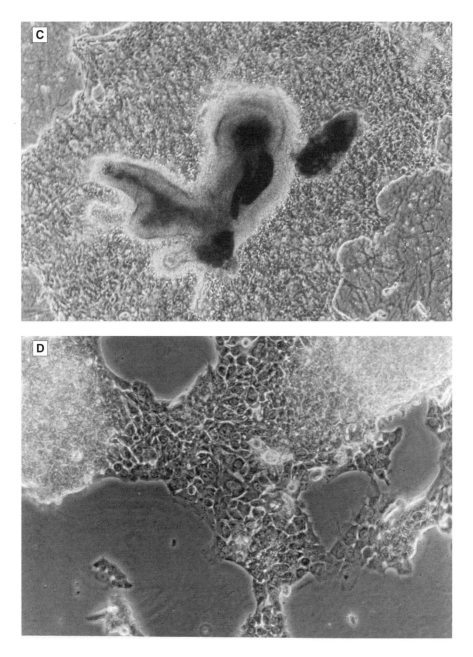

FIGURE 8.2 (continued)
of papillary TCC recapitulates the papillary morphology seen in (**D**). In contrast to HUC and papillary TCC, CIS show a loose, more irregular morphology consistent with a greater degree of cellular transformation, 40× (**E**) and 200× (**F**).

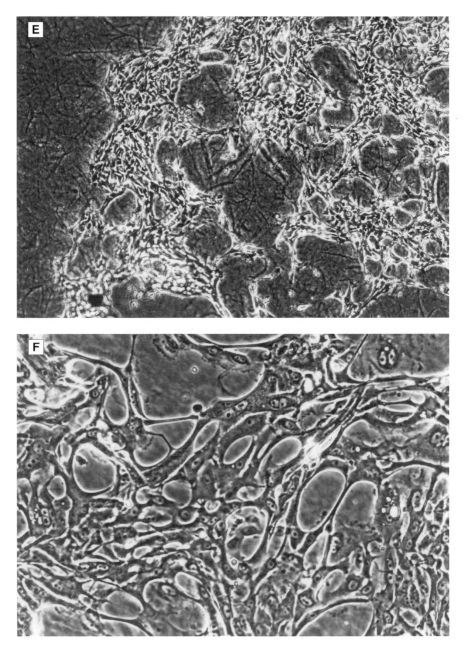

FIGURE 8.2 (continued)

30 to 50 population doublings;[13] therefore, some TCCs show extended lifespan in culture. In addition, infrequent TCC samples spontaneously immortalize, a phenomenon never observed in normal HUC cultures. Thus, extension of lifespan and spontaneous immortalization are parameters associated with the transformation of uroepithelium.

3. Anchorage-independent growth and tumorigenicity

Another useful endpoint of uroepithelial cell transformation is the acquisition of anchorage-independent growth.[10] To assess growth quantitatively requires about 1×10^6 viable cells distributed into 10 60-mm dishes assay. These types of cell numbers represent less than one confluent 100-mm TCC culture. The ability of cells to grow in agar without a solid substrate is a classic marker of cellular transformation in many systems.

Tumorigenicity is considered the ultimate endpoint of cell transformation and is an important parameter that can be measured with cultured cells. Typically, 5×10^6 cells per site are injected subcutaneously into athymic nude mice, two sites per mouse in duplicate mice per assay. This requires 2×10^7 actively growing cells collected from 6 to 8 100-mm dishes. Tumors vary greatly in both their latent period and growth rates (Figures 8.3A and B). A key advantage of studying the actively growing lines in these assays is that a tumor may contain a small subpopulation of cells that has the ability to progress. These more aggressive cells may be undetectable in a heterogeneous, uncultured tumor. The assays using cultured cells that we have described above may give the transformed minority a selective advantage, the population may be enriched for these more transformed cells, and their phenotype may thus be revealed *in vitro*.

B. Molecular Cytogenetic Studies

1. Comparative genomic hybridization and Southern analysis

The examination of uncultured tumors allows limited cytogenetic analysis. Detailed karyotyping is essentially impossible; however, tumor DNA can be collected for CGH analysis. CGH is an excellent starting point to scan the entire genome for gains and losses of chromosomal regions in a tumor sample.[14] Losses may be confirmed by conventional restriction fragment length polymorphism (RFLP) analysis or by polymerase chain reaction (PCR) assays for polymorphic markers on certain chromosomal regions.[15] Once regions of loss or gain are identified, the samples can then be tested for alterations in the TSGs or oncogenes located in these regions using Southern and northern analysis. CGH and RFLP each require approximately 10 μg of DNA per assay which can easily be obtained from a subconfluent 60-mm or 100-mm culture. A limitation of the techniques described in this section (CGH and RFLP) is that they detect clonal changes, while a subpopulation with a different, possibly more malignant, phenotype may go undetected.

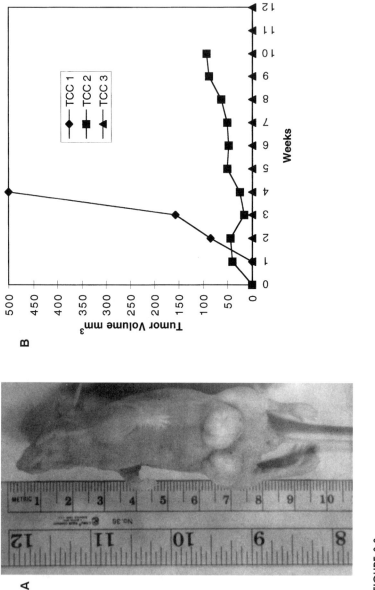

FIGURE 8.3

Formation of tumors in athymic nude mice after inoculation of cultured TCC: (**A**) Nude mouse with two very large subcutaneous tumors photographed at the time the animal was sacrificed and approximately 4 weeks after inoculation of cultured TCC. (**B**) Graph showing different growth rates in athymic nude mice after inoculation of three independent TCC lines. Note that TCC 1 formed a very aggressively growing tumor, while TCC 2 grew less aggressively, and TCC 3 did not form tumors.

2. Karyotypic and fluorescent *in situ* hybridization

The capability to culture TCCs generates metaphase cells for karyotyping and FISH (Figures 8.4A and B). Changes in ploidy, modal chromosome number, clonal chromosome region losses, gains, rearrangements, and marker chromosomes can be detected;[2,3,11,12,15-19] thus, karyotyping adds to and confirms CGH results. The use of FISH with whole chromosome paint (wcp) and specific probes is also useful for further analysis of rearrangements, deletions, or amplifications. Karyotyping and FISH can be done with the cells from one proliferating 60- or 100-mm culture dish.

C. Molecular Genetic Analyses

1. Genetic losses and alterations of tumor suppressor genes

With the advent of molecular genetics, tumor samples can now be tested for alterations in specific genes using Southern analysis, PCR technology, RFLP, etc. Such studies can and have been done with many samples of uncultured tumor materials. The *in vitro* system complements studies done with uncultured materials in several ways. First, cultured material may be less likely to contain normal cells. Second, culture may lead to selection *in vitro* for a minority of cells that are more malignant and will in time dominate the tumor *in vivo*.

The primary method for assessing protein expression in uncultured tumor samples is immunohistochemistry. In contrast, tumor cell cultures provide a rich source of protein for western blot analysis and immunoprecipitation studies. This is important for the assessment of altered gene activity. For example, a 60-mm culture dish provides sufficient protein for approximately 5 to 10 westerns using 25 to 50 µg of protein per lane.[20,21] Examples of proteins in which alterations can be detected using western analysis include CDKN2/p16, p53, pRb, ErbB2, and EGFR. For example, normal cells show low to undetectable p16, except at senescence (Figure 8.5). In contrast, tumor cells with altered pRb show increased levels of p16. In addition, either loss or stabilization of p53 suggests a *TP53* mutation (Figure 8.5). Western analysis is a good initial screening method to detect certain genetic alterations. An alternative approach to assay proteins is immunoprecipitation.[22] Immunoprecipitation typically involves radioactively labeling proteins in the cell and then using antibodies to precipitate the protein(s) of interest.

When alterations in expression suggest a gene is altered, it can be examined for mutations using standard molecular methods such as single-strand polymorphism analysis and sequencing.[23] Promoter DNA methylation of genes that are not expressed can also be examined when DNA mutations are not detected. Notably, some data suggest that methylation alterations[24,25] may occur frequently *in vitro*, which could be a problem for the interpretation of alterations found in cultured materials.

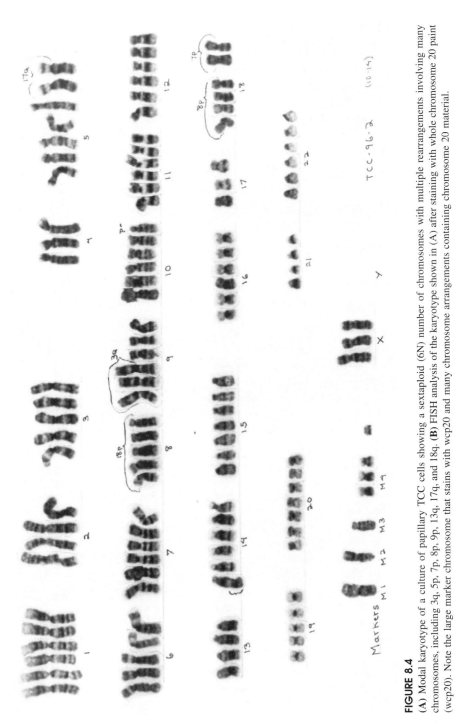

FIGURE 8.4

(A) Modal karyotype of a culture of papillary TCC cells showing a sextaploid (6N) number of chromosomes with multiple rearrangements involving many chromosomes, including 3q, 5p, 7p, 8p, 9p, 13q, 17q, and 18q. (B) FISH analysis of the karyotype shown in (A) after staining with whole chromosome 20 paint (wcp20). Note the large marker chromosome that stains with wcp20 and many chromosome arrangements containing chromosome 20 material.

B

FIGURE 8.4 (continued)

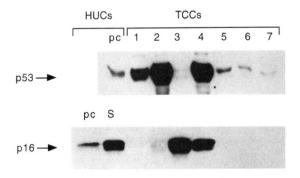

FIGURE 8.5
Western blot analysis for p16 and p53 in independent samples of cultured TCC. Note that the p53 protein band is strong in lanes 1, 2, and 4. Sequencing shows that p53 in lane 1 is a wild-type, apparently stabilized p53, while mutations were identified in TCCs in lanes 2 and 4. TCCs from lane 3 contained a splice mutation in *TP53*; p53 shows normal levels in other lanes. The level of p16 in normal precrisis (pc) HUCs is shown. The level of p16 is elevated in the senescent (S) HUC sample, as expected. Note the loss of p16 in lanes 1, 2, 5, 6, and 7.

2.   Genetic gains and activation of oncogenes

In addition to TSGs leading to a tumorigenic phenotype, oncogenes play an important role. One way a proto-oncogene can be activated to an oncogene is through mutation. The classic example of activation by a point mutation is the *RAS* proto-oncogene. A direct method to detect such a point mutation alteration is by sequencing. For example, *RAS* mutations typically occur in codons 12, 13, or 61.[26,27] The protein product of *RAS* acts in a cellular signaling pathway, being activated when the proper upstream signal is received and turned off when it is not present. Activated *ras* acts as an oncogene by continually sending a growth signal to the nucleus. Other mechanisms for oncogene activation are amplification and overexpression. Either alteration can be detected using western analysis for protein.

Comparative genomic hybridization detects amplified regions, putatively indicating the presence of an oncogene whose overexpression contributes to tumorigenesis. For example, *ERBB-2* on chromosome 17q and *EGFR* on 7p are often over represented in breast or bladder cancer.[28] Amplification of 20q is also often observed in breast and bladder cancers.[17] The value of *in vitro* cell cultures is that the biological significance of these amplifications can be correlated with phenotypes. Furthermore, the putative phenotypes of the altered gene(s) can be tested using chromosome or gene transfer.

3.   Mutator phenotype cancer genes

An additional class of genes that may be of importance in tumorigenicity and in progression are genes that cause genome instability.[2] Certain genes such as *hMLH-1* or *MSH-2* when altered or mutated can lead to greater instability in the genome due to their involvement in DNA repair processes. If these DNA repair genes are mutated and DNA damage occurs, the damage may not be corrected before the genome is replicated in the S phase, leading to the incorporation of additional mutations in the genome. The loss of the *hMLH-1* gene has been shown to lead to microsatellite instability throughout the genome. Following cytogenetic studies (karyotyping or CGH) and detection of considerable genomic instability in a sample, genes such as *hMLH-1* could be sequenced to determine if their alteration correlated with increased instability. Finally, PCR may be used to detect microsatellite instability in di- or tri-nucleotide repeat regions in the genome.[20] A high level of microsatellite instability could indicate a more progressed or transformed tumor phenotype.

4.   Telomerase activation and telomere stabilization

In addition to instability genes that may be involved in tumor progression, there are other markers that can be examined in tumor samples, such as telomerase activity and telomere length. Telomeres are the protective caps on the ends of chromosomes that shorten during cellular replication if the telomerase enzyme is not functioning. Once a cell population immortalizes, the telomeres stabilize putatively by affecting the telomerase enzyme activity (but how this happens is not currently known). Telomere length can be tested on a Southern blot using DNA digested with restriction enzymes that cut outside the telomere, such as *Hinf*1 and *Rsa*1. The Southern blots are then probed with a telomere-specific probe (human telomeres repeat the sequence TTAGGG).[29,30] In addition, growing cultures *in vitro* can be harvested and checked for the presence of the telomerase enzyme using the telomere repeat amplification protocol (TRAP) assay (only a small number of cells are necessary, $<1 \times 10^5$) at various time points. Most tumor biopsies have detectable levels of telomerase, while normal cells do not.

D.   Assessment of Tumor Response to Clinical Treatments

1.   Response to chemotherapy drugs and radiation therapy

It is difficult, and sometimes impossible, to predict whether a particular chemo-therapy or radiotherapy will be effective against an individual patient's tumor. Information that could be gained prior to treating the patient might save time and suffering for the patient. *In vitro* testing of drug toxicity against tumors with different phenotypes and genotypes may provide valuable information concerning those drugs that might be effective in a clinical setting. These studies using cultured cells are complementary to animal models.

With *in vitro* cultures, new drug therapies could be tested for cytotoxicity on TCCs with known differences in their genotype. Normal HUCs provide an essential control. The assays that could be used to determine the toxicity of the drugs include reduction of cell counts or colony forming ability.[6,31]

The *in vitro* system can also be used to test the effectiveness of radiotherapy against tumors with different genotypes. An endpoint to examine in drug response or radiotherapy studies is apoptosis, a form of programmed cell death. The apoptotic response can be studied using morphological endpoints (TUNEL assay) or by gel electrophoresis to detect the ladder resulting from DNA fragmentation.[32]

2. Biochemical studies on cellular response to drugs

The metabolism of new drugs can be assayed *in vitro* to determine toxicity in the human system. This will allow the testing of different genotypes, such as the N-acetyltransferase slow vs. fast phenotype on drug metabolism.[33] With these tests, for all practical purposes only the tumor lines can be used due to the large number of cells needed.

IV.   Summary

We have developed an *in vitro* tissue culture system for clinical TCC samples. The system allows the growth and expansion of primary TCC samples on a collagen gel substrate in a semidefined medium. Once the cells start to proliferate, they have a limited lifespan and most will senesce after 2 to 7 passages in culture, but occasionally TCC samples spontaneously immortalize and form cell lines. The ability to routinely culture TCC samples has allowed us to amplify the number of cells available from a primary tumor for testing in a number of different molecular, biological, or cytogenetic studies. This system may be invaluable for testing certain tumor parameters where proliferating cells or a large number of cells are required. Also, the system has suitable biological controls, as normal HUC also grow in culture. This system may permit the testing of different genetic or biological parameters in a tumor that may have translational relevance in the clinic for prognostic or diagnostic value. In addition, this may serve as a useful system for research into drug development or drug toxicity, or testing newly discovered candidate TSGs or oncogenes in the bladder cancer pathogenesis model. In the future, cell culture methods may contribute to answering questions with translational value. Examples of such questions are which papillary tumors are likely to progress and should be treated aggressively, and what are the best treatment strategies for tumors with specific genotypes/phenotypes.

Acknowledgments

We thank Dr. Lorraine F. Meisner and Mr. Charles Harris for their invaluable help with the cytogenetic FISH analysis. This work was supported by NIH R01CA29525-16 and NIH R01CA67158-02 to CAR.

References

1. Vogelzang, N. J., Scardino, P. T., Shipley, W. U., and Coffey, D. S., Eds., *Comprehensive Textbook of Genitourinary Oncology,* Williams & Wilkins, Baltimore, MD, 1996.

2. Reznikoff, C. A., Belair, C. D., Yeager, T. R., Savelieva, E., Blelloch, R. H., Puthenveetil, J. A., and Cuthill, S., *Semin. Oncol.,* 23, 571, 1996.

3. Reznikoff, C. A, Kao, C., Messing, E. M., Newton, M., and Swaminathan, S., *Semin. Cancer Biol.,* 4, 143, 1993.

4. Reznikoff, C. A., Loretz, L. J., Pesciotta, D. M., Oberley, T. D., and Ignjatovic, M. M., *J. Cell. Physiol.,* 131, 285, 1987.

5. Schmidt, W. W., Messing, E. M., and Resnikoff, C. A., *J. Urol.,* 132, 1262, 1984.

6. Loretz, L. J. and Reznikoff, C. A., *In Vitro Cell. Develop. Biol.,* 24, 333, 1988.

7. Reznikoff, C. A., Chen, X.-R., and Messing, E. M., in *Cell and Tissue Culture: Laboratory Procedures,* Doyle, A. and Griffiths, J. B., Eds., John Wiley & Sons, New York, 1997.

8. Reznikoff, C. A., Belair, C. D., Savelieva, E., Zhai, Y., Pfeifer, K., Yeager, T., Thompson, K. J., DeVries, S., Bindley, C., Newton, M. A., Sekhon, G., and Waldman, F., *Genes Develop.,* 8, 2227, 1994.

9. Belair, C. D., Blelloch, R. H., and Reznikoff, C. A., *Radiat. Oncol. Invest.,* 3, 368, 1996.

10. Reznikoff, C. A., Gilchrist, K. W., Norback, D. H., Cummings, K. B., Erturk, E., and Bryan, G. T., *Am. J. Pathol.,* 111, 263, 1983.

11. Christian, B. J., Loretz, L. J., Oberley, T. D., and Reznikoff, C. A., *Cancer Res.,* 47, 6066, 1987.

12. Kao, C., Wu, S.-Q., Bhattacharya, M., Meisner, L. F., and Reznikoff, C. A., *Genes, Chromosomes, Cancer,* 4, 158, 1992.

13. Harley, C. B., *J. NIH Res.,* 7, 64, 1995.

14. Kallioniemi, A., Kallioniemi, O.-P., Citro, G., DeVries, S., Kerschmann, R., Carroll, P., and Waldman, F., *Genes, Chromosomes, Cancer,* 12, 213, 1995.

15. Klingelhutz, A. J., Wu, S.-Q., Bookland, E. A., and Reznikoff, C. A., *Genes, Chromosomes, Cancer,* 3, 346, 1991.

16. Kao, C., Wu, S.-Q., DeVries, S., Reznikoff, W. S., Waldman, F. M., and Reznikoff, C. A., *Genes, Chromosomes, Cancer,* 8, 155, 1993.

17. Savelieva, E., Belair, C. D., Newton, M. A., DeVries, S., Gray, J. W., Waldman, F., and Reznikoff, C. A., *Oncogene,* 14, 551, 1997.

18. Meisner, L. F., Wu, S.-Q., Christian, B. J., and Reznikoff, C. A., *Cancer Res.,* 8, 3215, 1988.

19. Wu, S.-Q., Storer, B. E., Bookland, E. A., Klingelhutz, A. J., Gilchrist, K. W., Meisner, L. F., Oyasu, R., and Reznikoff, C. A., *Cancer Res.,* 51, 3323, 1991.

20. Yeager, T., Stadler, W., Belair, C., Puthenveettil, J., Olapade, O., and Reznikoff, C. A., *Cancer Res.,* 55, 493, 1995.

21. Reznikoff, C. A., Yeager, T. R., Belair, C., Savelieva, E., Puthenveettil, J. A., and Stadler, W. M., *Cancer Res.,* 56, 2886, 1996.

22. Kao, C., Huang, J., Wu, S.-Q., Hauser, P., and Reznikoff, C. A., *Carcinogenesis,* 14, 2297, 1993.

23. Fujimoto, K., Yamada, Y., Okajima, E., Kakizoe, T., Sasaki, H., Sugimura, T., and Terada, M., *Cancer Res.,* 52, 1393, 1992.

24. Merlo, A., Herman, J. G., Mao, L., Lee, D. J., Gabrielson, E., Burger, P. C., Baylin, S. B., and Sidransky, D., *Nature Med.,* 1, 686, 1995.

25. Gonzalez-Zulueta, M., Bender, C. M., Yang, A. S., Nguyen, T., Beart, R. W., Van Tornout, J. M., and Jones, P. A., *Cancer Res.,* 55, 4531, 1985.

26. Christian, B. J., Kao, C., Wu, S.-Q., Meisner, L. F., and Reznikoff, C. A., *Cancer Res.,* 50, 4779, 1990.

27. Pratt, C. I., Kao, C., Wu, S.-Q., Gilchrist, K. W., Oyasu, R., and Reznikoff, C. A., *Cancer Res.,* 52, 688, 1992.

28. Messing, E. M. and Reznikoff, C. A., *J. Cell. Biochem.,* 161(Suppl.), 56, 1992.

29. Klingelhutz, A. J., Barber, S. A., Smith, P. P., Dyer, K., and McDougall, J. K., *Mol. Cell. Biol.,* 14, 961, 1994.

30. Shay, J. W., *Molec. Med. Today,* 1, 376, 1995.

31. Reznikoff, C. A., Loreta, L. J., Johnson, M. D., and Swaminathan, S., *Carcinogenesis,* 7, 1625, 1986.

32. Puthenveettil, J. A., Frederickson, S. M., and Reznikoff, C. A., *Oncogene,* 13, 1123, 1996.

33. Pink, J. C., Messing, E. M., Reznikoff, C. A., Bryan, G. T., and Swaminathan, S., *Drug Metabol. Disposition,* 20, 559, 1992.

Section IV

Transgenes, Knockout Mice, and Mouse Models

Chapter **9**

Transgene Analysis in Mouse Embryonic Stem Cells Differentiating *In Vitro*

*Nicole Faust, Susanne König,
Constanze Bonifer, and Albrecht E. Sippel*

Contents

0-8493-4411-5/98/$0.00+$.50

I.   Introduction

With increasing research in the field of gene therapy it has become more and more important to generate appropriate gene expression vectors. Such vectors must be capable of driving expression of the introduced gene at a physiological level, ideally unaffected by genomic position effects, i.e., influences from the neighboring chromatin of the host genome at the random site of integration. Additionally, with systems in which the gene will be preferentially delivered to stem cells, e.g., the hematopoietic system, it must be guaranteed that the gene will only be activated in the correct cell type. It is therefore desirable to have model systems for testing expression vectors with respect to these features. Cell lines, representing a particular cell type, are insufficient models, since constructs active in such a cell line are not necessarily active when introduced into developing systems, where cell differentiation takes place.[1] Most likely, the reason for this is that exogenous gene constructs tend to integrate into active loci of the host genome with an "open" chromatin configuration. In cell lines this will facilitate transgene expression; however, if the cell undergoes differentiation after gene transfer, an active locus may be inactivated and adopt a condensed chromatin structure, impeding transgene activity. Taken together, this means that vectors that are to be used for stem cell therapy have to be tested by introducing them into stem cells before differentiation. One way to achieve this is the generation of transgenic animals.

A less involved alternative is provided by totipotent mouse embryonic stem (ES) cells differentiating *in vitro.* ES cells can be grown in their undifferentiated, totipotent state almost indefinitely. When induced to differentiate *in vitro,* they can, depending on the differentiation conditions, develop into a variety of cell types. These include epithelial cells, fibroblasts, hematopoietic cells, neural cells, and muscle cells.[2-4] Recently, a highly reproducible method for the induction of hematopoiesis has been developed.[5] With this method, ES cells form embryoid bodies (EBs) containing blood islands, which produce cells of different hematopoietic lineages. By adding certain cytokine combinations, the expansion of distinct lineages can be stimulated selectively.

Here, we will describe how this system can be used to analyze the capability of transgene constructs to autonomously govern cell-type-specific gene expression within the hematopoietic system. The emphasis of this chapter is on the myeloid lineage, for which we studied expression of a chicken lysozyme transgene; however, the methods presented here are easily adaptable for other hematopoietic or nonhematopoietic lineages.

II.   Experimental Strategy

The experimental strategy to analyze the capacity of an expression vector to direct cell-type-specific transgene expression, which is uninfluenced by position effects, is schematically outlined in Figure 9.1 First, the construct is integrated stably into the genome of ES cells which are subsequently induced to differentiate into embryoid bodies containing different hematopoietic cell types as well as nonhematopoietic cells. To obtain stable transfectants, ES cells are transfected with a plasmid carrying an antibiotic-resistance gene and the transgene construct. Cotransfection of two separate plasmids is preferable to linking the two genes on the same DNA molecule, as the proximity of an actively transcribed antibiotic-resistance gene might affect the regulation of the construct to be analyzed. Although the formation of mixed clusters after cotransfection of two constructs has been reported,[6] we have failed to detect interspersed antibiotic resistance plasmid sequences in the transgene construct clusters. After DNA transfection and antibiotic selection, resistant ES cell colonies can be isolated and transferred to 24-well plates. To save time and work, a rapid polymerase chain reaction (PCR) screen for transgene integration is performed (Figure 9.1). Only colonies that are positive for transgene integration in the PCR assay need be expanded further.

At this stage, aliquots of the clones are frozen for later experiments. In parallel, genomic DNA is prepared to confirm correct transgene integration and to determine the number of integrated copies in Southern blot analysis. Clones that have the transgene integrated without rearrangements are subjected to *in vitro* differentiation (IVD).

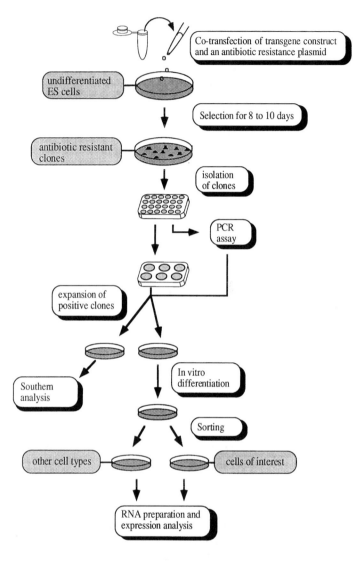

FIGURE 9.1

Schematic outline of the experimental procedures to analyze the activity of a transgene construct in the ES cell *in vitro* differentiation system.

To analyze cell-type specificity of transgene expression, the differentiated cells have to be sorted into fractions enriched or depleted for the cells of interest. Macrophages, for example, can be sorted easily by their selective plastic adherence.[7] For other cell types, density gradients or methods based on specific antibody affinity, such as fluorescence-activated or magnetic cell sorting, are available. Finally, RNA is prepared from different cell fractions, including undifferentiated ES cells, and analyzed for transgene expression. In addition to cell-type specificity, the level of expression has to be determined accurately for several independently derived clones. For these quantitative analyses, an endogenous marker gene will be useful, whose expression level presents a standard for the abundance of the specific cell type. For macrophages, the mouse lysozyme M gene can serve as such a standard.[7]

III. Generation of Stably Transfected Embryonic Stem Cell Clones

A. Routine Culture of Embryonic Stem Cells

For maintaining and differentiating ES cells, a well-equipped tissue culture facility and some prior tissue culture experience will be needed. Methods to culture embryonic stem cells have been described in great detail by others;[8] therefore, only a very short description of our routine culture conditions will be given here.

Embryonic stem cells of cell line CCE[9] are cultured in a humidified incubator at 37°C, 5% CO_2 in Dulbecco's modified eagle medium (DMEM, 4500 mg/l gluccose; Gibco-BRL) supplemented with 15% fetal calf serum (FCS, selected batch), 1.5×10^{-4} M monothioglycerol (MTG; Sigma), 50 U/ml penicillin, 50 mg/ml streptomycin, 2 mM glutamine, and 25 mM HEPES buffer (all Gibco-BRL). To maintain the undifferentiated state, ES cells are grown on gelatinized tissue culture dishes containing a feeder layer of mitomycin-treated STO fibroblasts.[8] In addition, the medium is supplemented with LIF (leukemia inhibitory factor), which is obtained as medium conditioned by CHO 8/24 720LIF-D(1) cells (Genetics Institute, Inc., MA) and is added to a final concentration of 0.7%. ES cells are seeded at a density of 2×10^4 cells per cm^2 and usually are subcultured at a ratio of 1:8 to 1:4 after 2 to 3 days when they are 50 to 75% confluent.

B. Stable Transfection of Embryonic Stem Cells

As pointed out above, the construct to be analyzed and the antibiotic-resistance gene, which is needed for selection of stable transfectants, should not be linked physically and hence have to be cotransfected. For successful cotransfection, the calcium phosphate coprecipitation method will work far better than electroporation. In general, calcium phosphate transfection yields relatively high copy numbers; however, the number of integrated copies can be controlled to a certain extent by transfecting the DNA either in linear or in circular form. Transfection of circular DNA leads to integration of only 1 to 5 copies per cell, whereas linearized DNA will form clusters of sometimes more than 100 copies (Table 9.1). Both the hygromycin-resistance gene and the neomycin-resistance gene will perform well as selection markers in ES cell transfection. With these resistance genes, hygromycin B (Boehringer Mannheim, working concentration 175 mg/ml) or G418 (Gibco-BRL, working concentration 600 µg/ml) are used as selecting agents, respectively.

TABLE 9.1
Variation of Copy Numbers after Transfecting ES Cells with Linear or Circular DNA by Calcium Phosphate Precipitation

Experiment[a]	Form of DNA	Transfection Efficiency[b] (%)	Clones Positive in PCR Assay[c]	Clones Positive in Southern[d]	Correct Integration Events[e]	Copy Number (mean; range)
1	Circular	0.047	n.d.	12	4	1.25; 1–2
2	Circular	0.025	n.d.	14	7	4.28; 1–16
3	Linear	0.039	n.d.	9	4	36.75; 2–100
4	Linear	0.024	n.d.	14	5	34.2; 1–70
5	Linear	0.025	9/22	9	5	123; 50–160
6	Linear	0.020	10/24	9	5	110; 25–250

[a] In experiments 1 to 4, ES cells were transfected with 30 µg of a single plasmid carrying the hygromycin resistance gene driven by the HSV-Tk promoter. In experiments 5 and 6, transgene constructs carrying the chicken lysozyme gene domain (30 µg) were cotransfected with 3 µg of the unlinked neomycin-resistance plasmid pMC1 neopolyA (Stratagene). All transfections were carried out using the Ca-phosphate transfection method as described in Section III.B.1.

[b] Number of resistant colonies after antibiotic selection divided by the number of transfected cells.

[c] ES cell clones were tested for integration of the chicken lysozyme transgene construct by PCR as described in Section III.C. The number of clones displaying a transgene-specific signal per number of analyzed clones is given (n.d.: not done).

[d] ES cell clones were tested for integration of the transgene constructs in Southern blots.

[e] Number of clones which carry the construct integrated without rearrangements as shown by Southern analysis.

The DNA to be transfected must be of high quality. DNA prepared with commercially available DNA purification kits from various suppliers contains a high proportion of "nicked" DNA. We thus recommend using DNA purified on CsCl-gradients. The ES cells should be of an early passage. With later passages, the risk of clonal variation increases, meaning that not all of the clones obtained after stable transfection will be able to differentiate property. If no early passage is available, subcloning of the cells and choosing a subclone with very good differentiation potential provide a solution.

1.   Calcium phosphate cotransfection

Transfections are carried out following the method described by Chen and Okayama.[10] Sixteen hours before transfection, plate ES cells onto gelatin-treated tissue culture plates *without* feeder cells at a density of 2.5×10^3 cells per cm^2 (one plate per construct plus one plate for a mock transfection without DNA). Ethanol-precipitate 20 to 50 µg of the plasmid to be transfected and equimolar amounts of pMClneopolyA (Stratagene) carrying the neo[r] gene, store DNA in ethanol at –70 to –20°C. Approximately 4 hr before transfection, aspirate medium from the ES cells, replace by fresh medium, and put cells back in the incubator. Centrifuge ethanol-precipitated DNA at 4°C and wash once with 70% ethanol. Remove 70% ethanol in the laminar flow hood and leave lids open to let the pellet dry. After drying, resuspend the pellet carefully in 450 µl H_2O (for example, by shaking for 1 hr at room temperature). To prepare precipitate, briefly spin tubes with dissolved DNA to collect contents at the bottom of the tube, and then, under the laminar flow hood, add 50 µl of 2.5 M $CaCl_2$ and mix by gently pipetting up and down. Add 500 µl of 2× BBS (50 mM BES; 280 mM NaCl; 1.5 mM Na_2HPO_4, pH 6.95) and mix gently by inversion of the tube. *Do not vortex.* Allow the tubes to stand for 10 to 20 min. Resuspend the formed precipitate by gently pipetting up and down, then add it dropwise to the cells, while swirling the plates. Incubate cells at 37°C, 5% CO_2, for 20 hr. By that time, crystals which are visible under the microscope will have formed. Remove crystals completely by aspirating the medium, washing the cells twice with PBS (140 mM NaCl; 2.5 mM KCl; 8.1 mM Na_2HPO_4; 1.5 mM KH_2PO_4, pH 7.5) and replacing with fresh medium. After 8 to 9 hr of further incubation, the cells are trypsinized and split at a ratio of 1:5 onto new plates containing a feeder layer of neomycin-resistant STO fibroblasts.[11] Twenty hours after splitting, G418 (Gibco-BRL) is added to the medium to a final concentration of 600 µg/ml. Medium is then exchanged daily and G418 is freshly added from a stock solution kept at –20°C. After 8 to 10 days, G418-resistant colonies appear. When all cells on the mock-transfected plates have died, the resistant colonies can be isolated as described below.

2.   Isolation of embryonic stem cell clones

Embryonic stem cell colonies should be picked once they are of sufficient size, which is approximately the diameter of the opening of a Gilson P20 tip. In no case should they become too large or even turn brown in their center. Since colony growth is not absolutely synchronous, cell isolation is done over a period of 3 to 4 days.

One day before isolation of the first colonies, 24-well plates are gelatin-treated, and feeder cells are plated at 5×10^4 cells per well. Place small drops (10 to 20 μl) of Trypsin/EDTA (0.25% trypsin [Sigma #T-8128], 0.04% EDTA in PBS) on *bacterial*-grade plastic dishes. Under the microscope, mark the positions of the colonies to be picked. This makes it easier to locate them during the picking process and, in addition, prevents accidental picking of cells from the same colony if the plate is re-used to isolate new colonies after 1 to 2 days. Under an inverted microscope, detach the colony from the plate with the tip of a Gilson P20 Pipetman set to 20 μl. The colony will come off as a solid aggregate and can be drawn into the tip as a whole. Try to transfer as little medium as possible, as the FCS will inhibit trypsin activity. Alternatively, the medium can be replaced by PBS during the picking process; however, we prefer to leave the cells in culture medium during that time, to improve their viability. Release the colony into one of the trypsin drops, cover the plate, and incubate for 3.5 min at 37°C. Then, create a single cell suspension by vigorously pipetting the solution up and down. To minimize contamination risk, this step should be done under the laminar flow hood. It can also be useful to use safety filter tips at this step. Check under the microscope that the colony has been disaggregated into single cells, then transfer the cells to one well of the 24-well plate containing feeder cells and 1 ml of ES cell culture medium with 600 μg/ml G418. Exchange medium every day. After 2 to 5 days, cells will be ready for transfer onto feeder layer-containing 6-well plates. The cells should be grown in G418-containing medium at least until this step; we recommend culturing them in the presence of G418 until they are frozen, to avoid carry-over of nontransfected cells.

C.   Identification of Clones Carrying the Transgene Construct

After cotransfection of the transgene construct and an unlinked neo[r] gene as described above, only about 50% of the G418-resistant colonies carry the transgene (Table 9.1). To identify transgene carrying clones as soon as possible, a rapid PCR assay can be performed during isolation or initial expansion of the clones. When the colonies are picked, half of their cells can be taken for quick DNA preparation and PCR assay at that time point. However, we found higher reproducibility if the PCR assay is done when more cells are available, i.e., during the transfer of cells from 24- to 6-well plates.

1.   Quick preparation of DNA from embryonic stem cells for polymerase chain reaction assay

The DNA isolation protocol described here is based on a method by Kim and Smithies.[12] Cells for DNA preparation are obtained during the trypsinization step.

After a single cell suspension in trypsin/EDTA has been created, transfer half of the cells to an appropriate plate (6- or 24-well) for further culturing. Transfer the other half (approximately 10 μl) to a microcentrifuge tube, add 20 μl of H_2O, and incubate at 100°C for 10 min. Spin briefly and add 1.5 μl proteinase K (Boehringer Mannheim; stock solution of 10 mg/ml). Digest for 1 hr at 50°C or overnight at 37°C. It is essential that the proteinase K be completely inactivated after this step. To achieve this, incubate the tube for *at least* 20 min at 100°C. DNA prepared by this method can be stored at −20°C. One third to one half of the preparation should be used in the PCR assay.

2. Polymerase chain reaction assay for transgene integration

Since the PCR assay is performed with very crude DNA preparations, a positive control should be included in the assay. Ideally, positive clones will be identified in a multiplex PCR with two primer pairs, one amplifying a fragment from genomic mouse DNA as internal control and one amplifying a transgene-specific fragment. The PCR conditions, such as $MgCl_2$ concentration, primer concentration, or annealing temperature, have to be established for each combination of primers and primer pairs. The manyfold aspects of optimizing PCR conditions cannot be considered here; the reader is referred to volumes dealing extensively with these problems.[13-15] If the selected primer pairs do not work in a multiplex PCR, try different combinations or alternatively, divide the sample and set up two separate reactions. The protocol for a typical reaction using primers MLP1 and MLP2 (Table 9.2) amplifying a fragment from the endogenous mouse lysozyme gene and the transgene-specific primers LysP1 and LysP2 (Table 9.2) specific for the transfected chicken lysozyme gene is given below.

Mix 10 μl template DNA (prepared as described above), 3 μl of 10× PCR buffer (500 m*M* KCl; 100 m*M* Tris-HCl, pH 8.3; 15 m*M* $MgCl_2$; 0.1% gelatin, 75 bloom, Sigma; 2 m*M* each dNTP), and 10 pmol of each primer (MLP1, MLP2, LysP1, LysP2) in a volume of 30 μl. Heat to 94°C for 3 min, spin briefly, and chill on ice. Add 1 U of Taq-polymerase (Gibco-BRL) and mix thoroughly. Cover with a thin layer of paraffin oil (oil can be omitted if a thermocycler with a heated lid is used). Amplify DNA in a thermocycler by performing 30 to 35 cycles of 30-sec denaturation at 94°C, 30-sec annealing at 60°C, and 1-min elongation at 72°C. After the last cycle, analyze half of the PCR product by standard gel electrophoresis.

Although the PCR assay is very reliable in identifying transgene-carrying clones (Table 9.1), it cannot tell whether the construct has been integrated without rearrangements, as only a relatively small fragment is amplified. On the other hand, we have found that the transgene construct is rearranged in a significant number of ES cell clones (Table 9.1); therefore, correct transgene integration has to be verified in Southern blot experiments,[16] which can simultaneously serve to determine the number of integrated transgene copies.

TABLE 9.2
Primers for PCR Amplification of Mouse Lysozyme (mlys)- or
Chicken Lysozyme (clys)-Specific Fragments

PCR	Primers	Amplified Fragment (bp)	PCR Conditions		
			MgCl$_2$ (mM)	Primer Concentration (µM)	Annealing Temperature (°C)
clys (genomic DNA)	LysP1: 5' AGGGGCAGGAAGCAAAGCAC 3' LysP2: 5' CCACCTGCCACTGAATGGCT 3'	569	1.5	0.3	60
mlys (genomic DNA)	MLP1: 5' TCGGCCAGGCTGACTCCATA 3' MLP2: 5' ACCCAGCCTCCAGTCACCAT 3'	152	1.5	0.3	60
mlys (RT-PCR)	MLP2: 5' ACCCAGCCTCCAGTCACCAT 3' MLP3: 5' CAGTGCTTTGGTCTCCACGG 3' (MLP$_{comp}$: 5' cagtgctttggtctccacggCGTGCTGAGCTAAAACACACC 3')	228 (194)	1.5	0.15	65
clys (RT-PCR)	LysEx3: 5' GATCGTCAGCGATGGAAACGGC 3' LysEx4: 5' CTCACAGCCGGCAGCCTCTGAT 3' (LysEx$_{comp}$: 5' ctcacagccggcagcctctgatCTTGCAGCGGTTGCGCCA 3')	101 (80)	1.0	0.15	65

IV. *In Vitro* Differentiation of Embryonic Stem Cells

A.   General Considerations

A variety of methods to induce ES cell differentiation have been reported. To generate nonlymphoid hematopoietic cells, the method of choice is the one described by Wiles and Keller.[5] This method is based on culturing the cells in semisolid methylcellulose medium and reproducibly produces embryoid bodies containing blood islands. A very detailed description of this method has been published by M. V. Wiles.[17] Here, only a short description of how to induce differentiation under specific conditions promoting the development of macrophages will be given. Some points are very critical to successful induction of hematopoiesis. These are water and medium quality, FCS quality, the general state of the ES cells, and the density at which the cells are seeded.

1.   Selecting reagents

To guarantee high-quality medium, the medium should be bought in liquid form. If the addition of water is necessary, as for preparing methylcellulose stocks, either ultrapure water should be used or tissue culture grade water should be bought. Development of hematopoietic cells during EB differentiation strongly depends on the batch of FCS used. About one in two batches will induce hematopoiesis, and one in five will give very good results. To test FCS batches, set up IVDs as described below and record plating efficiency, size of EBs, and percentage of EBs developing blood islands. Blood island formation can easily be scored using darkfield microscopy at low magnification. Blood islands will be visible by their red color. Additionally, development of the cell type of interest, e.g., macrophages, should be monitored.

2.   Selecting cells

 Embryonic stem cells used for IVDs should be growing well in compact colonies and displaying no signs of spontaneous differentiation. If the cells do not look healthy, use an earlier transfer generation. Should this not be possible, the state of the cells can often be improved by removing old feeders and spontaneously differentiated ES cells through preplating and growing the cells for one or two passages under routine culture conditions before setting up differentiations.

   For preplating, trypsinize the cells, add medium, and prepare a single cell suspension. Transfer the cells to a nongelatin-treated tissue culture plate and incubate them for 1 hr. By that time, feeder cells and ES cells which are no longer in the undifferentiated state will become adherent. Transfer the supernatant containing the undifferentiated ES cells to a fresh gelatinized, feeder layer-containing tissue culture flask.

   The density at which the EBs grow in methylcellulose cultures has proven to be very critical for good differentiation results. If the density is too high, the nutrients will be used up too quickly. Fresh medium can be added to a certain extent, but this will dilute the methylcellulose. On the other hand, if the density of EBs is too low, they will develop poorly and hematopoiesis is less efficient. This phenomenon is probably explained by growth factors being secreted by the developing EBs. Optimally, EB density should be 25 to 50 EBs per ml. Plating efficiency can vary considerably from clone to clone; therefore, it is necessary to determine the optimal ES cell number to be seeded for each clone.

B.   *In Vitro* Differentiation Under Macrophage Growth-Promoting Conditions

To induce ES cells to form EBs containing blood islands, IVDs are performed in methylcellulose media as described.[5,17] To stimulate selectively macrophage development, a combination of interleukin-1 (IL-1), IL-3, and macrophage-colony stimulating factor (M-CSF) is added to the cultures. Methylcellulose medium is commercially available (Terry Fox Laboratory, Vancouver). A cheaper alternative is to prepare a 1.8% methylcellulose stock as described[17] and to store it in aliquots at –20°C. Recombinant IL-1 can be obtained from Hoffmann-La Roche, Nutley. IL-3 can be provided in the form of medium conditioned by a producer cell line, and M-CSF is provided as L-cell-conditioned medium (LCM). For all cytokines added as conditioned media, it is advisable to prepare large batches, test, titrate them, and then store them in appropriately sized aliquots at –20°C.

1.   Preparation of IL-3-containing medium

Expand X63 Ag8-653 myeloma cells carrying an IL-3-expression plasmid[18] in DMEM supplemented with 5% FCS, 50 U/ml penicillin, 50 mg/ml streptomycin, 2 mM glutamine, 25 mM HEPES-buffer, and 1 mg/ml G418 to prevent loss of the expression plasmid. Use 200 ml of cell suspension (10^6 cells per ml) and centrifuge for 4 min at 200 g. Wash pellets three times in PBS. Then resuspend the cells in a total volume of 500 ml DMEM with 2% FCS, 50 U/ml penicillin, 50 mg/ml streptomycin, 2 mM glutamine, and 25 mM HEPES buffer *without* G418, and incubate for several days until the medium turns yellow. To harvest, centrifuge cells and then collect, filter-sterilize, and pool the supernatants into one batch. The IL-3 activity can be titrated by setting up IVDs using a range of concentrations of the conditioned medium and monitoring macrophage development. Usually, we find that 5% of IL-3 medium which was prepared in the way described above yields very good results.

2.   Preparation of L-cell-conditioned medium

Culture Ltk⁻ cells in tissue culture flasks in DMEM with 5% FCS, 50 U/ml penicillin, 50 mg/ml streptomycin, 2 mM glutamine, and 25 mM HEPES buffer until they are confluent. Aspirate medium, wash once with PBS, and add Iscove's modified Dulbecco's medium (IMDM, Gibco-BRL) with 2% FCS, 50 U/ml peni-cillin, 50 mg/ml streptomycin, and 2 mM glutamine (0.1 ml of medium per cm^2). Incubate for 3 days, harvest, and replace medium. One confluent cell layer can be used up to three times to condition medium. Centrifuge harvested medium to remove cell debris and collect it at 4°C. After the last medium harvest, pool all LCM into one batch and filter-sterilize. M-CSF activity can be titrated by setting up IVDs using a range of concentrations of the LCM and monitoring macrophage develop-ment. Usually, 10% of LCM, which was prepared in the above described way, yields very good results.

3.   *In vitro* differentiation protocol

In vitro differentiations should be set up in 3.5- or 6-cm bacterial grade plastic dishes. With larger plates, the EBs tend to accumulate at the rim, leading to an uneven distribution. For a 3.5- or 6-cm plate, 1.25 ml or 4 ml of differentiation medium will be needed, respectively.

To prepare 20 ml of differentiation medium, mix in a 50-ml centrifuge tube 20 ml of methylcellulose stock (1.8% methylcellulose in IMDM), 8 ml of IMDM (supple-mented with 50 U/ml penicillin, 50 mg/ml streptomycin, and 2 mM glutamine), 2 ml of FCS (preselected batch), 20 μl insulin (Sigma #I-6634; working stock 10 mg/ml), and 60 μl of freshly prepared monothioglycerol (MTG) stock (12.5 μl monothioglycerol [Sigma #M-6145] in 1 ml IMDM). As the methylcellulose stock is very viscous, it cannot be pipetted accurately. Either use a syringe to transfer the desired volume or just pour it into the centrifuge tube using the scale on the tube to determine the volume. The differentiation medium can be prepared several days before IVDs are set up and can be stored at 4°C; however, MTG should be added freshly. To specifically enhance development of a certain lineage, the appropriate cytokines have to be added. To promote macrophage development, add IL-1 at 100 U/ml, IL-3 (5% of conditioned medium), and M-CSF (10% of LCM). Mix the components very well by inverting the tube at least 10 times.

Trypsinize ES cells and create a single cell suspension in medium that is identical to the medium used for routine culture but does not contain LIF. Count cells carefully. Add cells to differentiation medium at the concentration shown to yield the desired density of EBs. If the plating efficiency of the cells is still unknown, start with different concentrations between 10^3 and 5×10^3 cells per ml. Note that plating efficiency increases with the density of plated cells. Again, mix very well by inversion and allow the tube to stand for 1 min. Finally, distribute the mix to the dishes. To release accurately defined amounts of medium, use a syringe with an 18-gauge needle; however, with most applications it is sufficient to use glass or plastic pipettes. The plates should be placed into larger plates with loosely fitting lids (for example, larger tissue culture plates) to prevent them from drying out but allowing CO_2 and oxygen exchange. Place plates into the incubator and do not move them for the first 3 days. The incubator should be opened as infrequently as possible during that time. After 3 days, EBs can be seen and counted under the microscope. Globinization usually becomes visible by day 8 and macrophages moving out of the embryoid bodies can be detected from day 12 onwards.

To promote development of other hematopoietic cells, different cytokine combinations can be used. In the presence of IL-3 alone, mast cells develop preferentially; alternatively, the addition of erythropoietin leads to increased numbers of erythrocytes.[5] B or T lymphocytes can be obtained with different differentiation protocols involving coculture of ES cells with stromal cells from osteopetrotic mice[19] or differentiation in a low-oxygen atmosphere,[20] respectively.

V.   Purification of Macrophages

A.   Macrophage Enrichment by Selective Adherence

To test whether the transgene construct is expressed cell type-specifically, the differentiating cells must be sorted into two fractions, enriched or depleted, for the cell type of interest. For macrophages this is easily done, since, in contrast to the other cell types developing during ES cell differentiation, macrophages are adherent to bacterial-grade plastic; however, they hardly adhere when methylcellulose is present. Thus, macrophages can be enriched by transferring the differentiation cultures from semi-solid methylcellulose to liquid suspension culture on bacterial dishes.

Some bacterial-grade dishes are very sticky and other cell types will adhere to them, also; therefore, the plates have to be tested for selective macrophage adherence in advance. In our hands, 6-cm plates obtained from Greiner (#633102) have been very efficient. This method yields nearly pure macrophage populations,[7] and it can be applied at a very large scale.

Between day 10 and 14 of differentiation, harvest EBs by adding 4 ml of PBS per 6-cm plate and gently swirling the plate. With a 10-ml pipette, transfer the diluted methylcellulose containing the EBs to a centrifuge tube. Rinse plate once with 4 ml of PBS. Centrifuge at 200 g, aspirate supernatant, wash pellet once with PBS, and resuspend in an appropriate volume of IMDM supplemented with 10% FCS (same batch as for IVD), 50 U/ml penicillin, 50 mg/ml streptomycin, 2 mM glutamine, 100 U/ml IL-1, 1.5% IL-3, and 10% LCM. Transfer the cells to bacterial-grade plastic dishes. Depending on the EB density, one plate of methylcellulose cultures should be transferred to 1 to 4 new plates. After about 10 days, a nearly confluent layer of macrophages can be obtained. To save the nonadherent fraction, transfer supernatant to new plates after 5 days.

B.   Other Sorting Methods

For cells that cannot be purified by selective adherence, other methods utilizing cell-type-specific antibodies can be applied. The most widely used method is fluorescence-activated cell sorting (FACS); however, FACS requires extremely expensive equipment and extensive training. An alternative is provided by magnetic cell sorting.[21] It is less complicated to establish this method, which usually produces cell populations of sufficient purity. In addition, magnetic cell sorting offers the advantage that significantly larger (several orders of magnitude) cell numbers can be sorted. It can thus be used to enrich for extremely rare cells that subsequently can be sorted by FACS. The method employs columns that are magnetizable and a permanent magnet producing a strong magnetic field. The cells to be enriched are stained with a cell-type-specific antibody and a secondary antibody which is coupled to small superparamagnetic microparticles (about 100-nm diameter). The mixture of magnetic antibody-bound and nonbound cells is then passed over the column within the magnetic field, retaining only the stained cells, which are later eluted outside the field. If sorting is done under sterile conditions, cells can be cultured again after magnetic separation. A magnetic cell sorter can be obtained from Miltenyi Biotec, Inc. (Bergisch Gladbach, Germany; Auburn, CA). The methods involved in magnetic cell sorting are described very accurately in the manual; however, if possible, the method should be learned in a laboratory where it is well established.

1.   Purification of embryo stem cell-derived macrophages by magnetic cell sorting

For the isolation of macrophages, the mouse macrophage-specific monoclonal antibody F4/80[22] can be used as primary antibody. Sterilize separation column (size A2, Miltenyi Biotec) by filling it with 70% ethanol, then wash twice with PBS and finally with PBS/0.5% BSA. Leave the column filled with PBS/0.5% BSA while preparing the cells. To prepare a single-cell suspension from EBs, they are digested with collagenase as described:[17] Between days 12 and 14 of differentiation collect EBs from at least 20 6-cm plates by PBS flush and centrifugation as described in Section V.A. Wash pellet once with PBS and resuspend in 10 to 20 volumes of collagenase solution (0.25% collagenase [Sigma #C-0130], 20% FCS [no special batch required] in PBS). Incubate with constant agitation at 37°C for 30 min. Add two volumes of PBS, pellet the cells, and resuspend in 5 ml of PBS. To obtain single cells, pass the suspension eight times through a 20-gauge needle. Count the cells and pellet 1×10^7 cells by centrifugation at 200 g for 5 min. Dilute primary antibody (F4/80, Serotec #MCA 497) 1:5 in PBS/3% BSA. Resuspend cell pellet in 100 µl of diluted primary antibody and incubate for 15 min at 6 to 8°C. To wash, add 1 ml of PBS/0.5% BSA, mix gently, and pellet cells at 600 g and 4°C for 5 min. Take off supernatant and resuspend cells in 80 ml of PBS/3% BSA. Add 20 µl of undiluted magnetic bead-conjugated secondary antibody (goat-anti-rat IgG, Miltenyi Biotec #485-01) and mix gently. To be able to check the separation efficiency, a fluorochrome-conjugated secondary antibody, (e.g., goat-anti-rat IgG, Cy3-conjugated, Dianova #112-165-003) can be added at this step. After a 15 min incubation at 6 to 8°C, add 1 ml of PBS/0.5% BSA and centrifuge at 600 g and 4°C for 5 min. Resuspend pellet in 500 µl of PBS/0.5% BSA. To cool the column, rinse it with 5 ml of ice-cold PBS/0.5% BSA. Place column into magnetic field and add cell suspension to the top. Let cells run through column with a 23-gauge needle fixed to the lower end of the column, controlling flow speed. Collect the nonretained cells as F4/80⁻ fraction. Rinse with 3 ml of PBS/0.5% BSA. To improve purification, remove column from magnetic field and press 5 ml of PBS/0.5% BSA from bottom to top through the column using a syringe attached to the side of the three-way stopcock at the bottom of the column. Place column back into magnetic field and let the wash solution run through. Rinse with 3 ml of PBS/0.5% BSA. Finally, collect F4/80+ cells by removing the column from the magnetic field, removing the needle from the outlet and pushing 4 ml of PBS/0.5% BSA through the column into a centrifuge tube. Centrifuge collected cells, resuspend them in media, and plate them for further culturing. After several days, when the cells have recovered from the separation procedure, RNA for expression analysis can be prepared.

VI.   Quantitative Expression Analysis

To examine gene expression, total cellular or polyA+ RNA can be analyzed for specific transcripts in northern blots, RNase protection assays, S1 protection analysis, or reverse transcription PCR (RT-PCR). Only the last method, which is universally applicable and is especially useful if material is limited, will be described here.

To perform quantitative expression analysis, a standard is needed which accounts for variations in RNA amounts, reverse transcription efficiency, and cDNA quality. Ubiquitously expressed housekeeping genes such as α-tubulin or β-actin can serve as such a standard; however, when transgene activity is to be measured in a particular cell type, a gene, which is known to be expressed constitutively but specifically in this cell type, can be used. This cell-type-specific standard offers the advantage that it normalizes not only for cDNA amount and quality but also for the abundance of the cells of interest. For macrophages, the mouse lysozyme M gene has been found a reliable cell-type-specific standard.[7] For other lineages, we recommend analyzing whether an endogenous, cell-type-specific gene can be used similarly. Once a cell-type-specific standard has been established, cells do not always have to be purified, but transgene expression can also be analyzed in mixed populations.

We will here describe a quantitative RT-PCR application to analyze mouse lysozyme (mlys) expression as a macrophage standard and we will also provide the conditions used to analyze expression of a chicken lysozyme transgene. The methods presented can be adapted easily for other endogenous genes or transgenes by selecting suitable primer pairs.

A.   RNA Isolation and Reverse Transcription

1.   RNA isolation

For isolation of total cellular RNA, standard methods can be used. The acid guanidinium-thiocyanate procedure described by Chomczynski and Sacchi[23] has been used successfully to prepare RNA from ES cells, EBs, and other cell types developing *in vitro* and is suitable for preparing RNA from small samples. Purified RNA should be dissolved in an appropriate volume of RNase-free water, and its quality should be checked by gel electrophoresis (the 28S rRNA band should have about twice the intensity of the 18S band). If RNA is present in sufficient amounts, its concentration can be determined accurately by spectrophotometry. Smaller RNA amounts can be estimated by comparing the signals to known standards on ethidium bromide stained gels.

2.   Reverse transcription

In the first step of RT-PCR (Figure 9.2B), single-stranded cDNA (ss cDNA) has to be produced by reverse transcription: Mix 1 μg of purified total cellular RNA with 0.5 μg of primer-oligonucleotide (oligo (dT)$_{12-18}$ or random hexamers) and add water to a final volume of 12.5 μl. Heat to 70°C for 10 min; afterwards cool immediately on ice. Centrifuge briefly and add 4 μl of 5× first-strand buffer (Gibco-BRL, supplied with reverse transcriptase), 2 μl 0.1 M DTT, and 0.5 μl polymerization mix (20 mM each dNTP; Pharmacia #27-2035-01). Vortex gently and collect contents of tube by short centrifugation. Put tube into 37°C thermoblock. After 2 min, add 1 μl (200 U) of reverse transcriptase (SuperScript®, Gibco-BRL #18053-017), mix by pipetting up and down gently, and incubate at 37°C for 1 hr. Stop the reaction by heat inactivation of the enzyme at 70°C for 30 min.

Single-stranded cDNA prepared as described above can be used directly for PCR. If an additional purification step is desired, purification can be done without significant product loss by using the QIAquick PCR Purification Kit (Qiagen) which was originally designed for purification of PCR products.

B.   Quantitative Reverse Transcription-Polymerase Chain Reaction

It is far beyond the scope of this article to discuss RT-PCR applications extensively. The reader is referred to volumes focusing on this subject.[13,14,24] Primer design is very critical for good RT-PCR results. Apart from general criteria, such as GC content, lack of palindromic sequences, and lack of complementary sequences between primers of one pair, RT-PCR requires that the primers reside in different exons. This enables a distinction between fragments amplified from cDNA and fragments amplified from contaminating genomic DNA (Figure 9.2A).

If possible, the primers should amplify a fragment that can be identified unambiguously by analytical restriction digests to ensure its specificity. Last, but not least, PCR with the selected primer pair should be sensitive: a clear signal should be obtained with 0.01 pg of plasmid DNA, once the amplification conditions have been optimized.

Due to reaction kinetics of PCR and extreme sensitivity of the reaction to minute variations in starting or amplifying conditions, it is very difficult to obtain reliable quantitative results. One way to circumvent these problems is the addition of a competitor fragment, which is amplified by the same primer pair but results in a fragment of slightly different size. The idea is to perform several parallel PCR reactions from one sample adding different amounts of the competitor fragment.

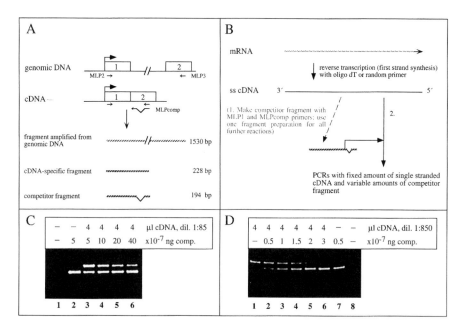

FIGURE 9.2

Quantitative analysis of mouse lysozyme mRNA expression by competitive RT-PCR. (**A**) Position of primers MLP2+3, which amplify a mouse lysozyme-specific fragment, and MLP$_{comp}$, which is used to generate a competitor fragment, and the resulting PCR fragments are depicted (for primer sequences see Table 9.2). (**B**) Experimental outline for RT-PCR. The competitor fragment is generated in advance from a single-stranded cDNA preparation, as shown here, or from a cDNA-containing plasmid. (**C**) Total cellular RNA was prepared from cells differentiated *in vitro* for 24 days under macrophage growth-promoting conditions. RNA (1 µg) was reverse transcribed; the product was purified with the QIAquick PCR purification kit (Qiagen), and resuspended in 50 µl H$_2$O. An aliquot of this single-stranded cDNA preparation was diluted 1:85. Competitive PCRs were set up by mixing 4 µl of the template with variable amounts of competitor fragment (4×10^{-6} to 5×10^{-7} ng) as indicated. A negative control reaction without template or competitor fragment (lane 1) and a reaction with competitor fragment only (lane 2) were also performed. (**D**) After the first round of PCR a second series of reactions was set up, using a 1:850 dilution of the ss cDNA preparation. This time the competitor fragment quantity was within the range of 5×10^{-8} to 3×10^{-7} ng. Lane 1: no competitor; lane 7: 5×10^{-8} ng competitor only; lane 8: negative control.

Only when both templates are present in equal concentrations will the intensity of the two bands be identical on ethidium bromide-stained gels (see Figure 9.2). Additional considerations on competitive PCR can be found in a recent article by Souazé et al.[25]

1.   Preparation of competitor fragments

A suitable competitor fragment can be prepared by constructing a template with an insertion or a deletion and performing PCR with the primer pair that will be used for the analysis. However, it is faster and more convenient, to use a method described by Celi et al.[26] which employs a third PCR primer (Figure 9.2A and Table 9.2). This third primer covers one of the two gene-specific primers and contains additional bases of the template (about 20 nucleotides), omitting a number of nucleotides. If this primer is used instead of the original primer, PCR results in amplification of a fragment that is shorter than the analytic PCR product (Figure 9.2A). The competitor fragment can be generated by PCR amplification from plasmid DNA containing a cDNA template. If no such template is available, it can also be generated from single-stranded cDNA. The fragment should be prepared in sufficient quantity to perform all planned experiments, guaranteeing that exactly the same amounts of competitor fragment are used between experiments. In addition, this will allow determination of concentration accurately by spectrophotometry. A competitor fragment for mouse lysozyme-specific RT-PCR can be generated using the primers MLP2 and MLPcomp (Table 9.2).

To prepare the fragment, mix the following in one PCR reaction tube: 1 μl of single-stranded cDNA (prepared from mouse macrophage RNA, corresponding to 1/50 of cDNA obtained from 1 μg RNA), 15 pmol or primer MLP2, 15 pmol of primer MLP_{comp}, 10 μl of 10× PCR buffer (500 mM KCl; 100 mM Tris-HCl, pH 8.3; 15 mM MgCl$_2$; 0.1% gelatin [75 bloom, Sigma], 2 mM each dNTP), 10 U of Taq-polymerase (Gibco-BRL), and water to a final volume of 100 μl. To prepare enough fragment, set up several parallel reactions. Amplify fragment by performing 35 cycles of 30-sec denaturation at 94°C, 30-sec annealing at 60°C, and 1-min elongation at 72°C. After the last cycle, add 10 min of elongation at 72°C, electrophorese PCR products on 2% agarose gel, excise fragment from the gel, and purify by standard elution methods.

2. Competitive polymerase chain reaction

In a first series of experiments, a range of dilutions of each cDNA preparation should be amplified with the specific primer pair in the absence of competitor fragment. Dilutions resulting in amplification which is within the linear range of the reaction should then be used for the competitive PCR. The next step will then be to set up PCRs with cDNA plus a competitor fragment, whereby a range of tenfold dilutions of the competitor fragment is used to get an idea of the order of magnitude of required competitor amounts. Further experiments are then performed with increasingly smaller ranges of competitor fragment concentrations (Figure 9.2C and D) until the amount of cDNA present in the sample has been determined with sufficient accuracy.

The protocol below gives an example for a competitive PCR for mouse lysozyme. Competitive RT-PCRs for chicken lysozyme transgene transcripts have been performed in similar experiments, using the primer pair LysEx3 and LysEx4 (Table 9.2) and a competitor fragment amplified with LysEx3 and $LysEx_{comp}$. To avoid handling of small volumes and to guarantee equal distribution, cDNA and competitor fragments should be added in volumes no smaller than 5 μl. Usually, the reactions will be done for about five cDNAs simultaneously.

Mix appropriately diluted cDNA and water to a final volume of 5 μl per reaction, heat for 5 min at 94°C, spin briefly, and distribute to reaction tubes. For each of the different competitor amounts to be added, multiply this quantity with the number of different cDNA samples, transfer this amount into a tube, and add water to a final volume of 5 µl per reaction. Heat for 5 min at 94°C, spin briefly, and distribute to the correct tubes. Prepare amplification mix containing 5 pmol of each primer (MLP2 and MLP3, Table 9.2), 3 µl of 10× PCR buffer (500 mM KCl; 100 mM Tris-HCl, pH 8.3; 15 mM MgCl$_2$; 0.1% gelatin; 2 mM each dNTP), and 2 U of Taq-polymerase in a total volume of 20 ml per reaction. Mix well and add 20 µl of amplification mix to each PCR tube. Mix well again and cover with paraffin oil if the thermocycler does not possess a heated lid. Amplify DNA by performing 30 cycles of 30-sec denaturation at 94°C, 30-sec annealing at 60°C, and 1-min elongation at 70°C. Analyze half of the PCR product on 6% polyacrylamide gels. The result of such an experiment with one cDNA preparation is shown in Figure 9.2D.

Abbreviations

BSA	Bovine serum albumin
DMEM	Dulbecco's modified Eagle medium
EB	Embryoid body
FACS	Fluorescence-activated cell sorting
FCS	Fetal calf serum
IL	Interleukin
IMDM	Iscove's modified Dulbecco's medium
IVD	*In vitro* differentiation
LCM	L-cell-conditioned medium
LIF	Leukemia inhibitory factor
M-CSF	Macrophage-colony stimulating factor
MTG	Monothioglycerol
PCR	Polymerase chain reaction
RT-PCR	Reverse transcription PCR
ss	Single-stranded

Acknowledgment

Ongoing research with transgenes in mouse ES cells differentiating *in vitro* is supported by grants to A.E.S. from the Deutsche Forschungsgemeinschaft (SFB 364, A2), The Fonds der Chemischen Industrie, and the German Israeli Foundation for Research and Development (G.I.F.).

References

1. Guy, L. G., Kothary, R., Derepentigny, Y., Delvoye, N., Ellis, J., and Wall, L, *EMBO J.,* 15, 3713, 1996.

2. Doetschman, T. C., Eistetter, H., Katz, M., Schmidt, W., and Kemler, R., *J. Embryol. Exp. Morphol.,* 87, 27, 1985.

3. Wobus, A. M., Wallukat, G., and Heschler, J., *Differentiation*, 48, 173, 1991.

4. Fraichard, A., Chassande, O., Bilbaut, G., Dehay, C., Savatier, P., and Samarut, J., *J. Cell Sci.*, 108, 3181, 1995.

5. Wiles, M.V. and Keller, G., *Development*, 111, 259, 1991.

6. Perucho, M., Hanahan, D., and Wigler, M., *Cell*, 22, 309, 1980.

7. Faust, N., Bonifer, C., Wiles, M. V., and Sippel, A. E., *DNA Cell Biol.*, 13, 901, 1994.

8. Robertson, E. J., in *Teratocarcinomas and Embryonic Stem Cells — A Practical Approach*, Robertson, E. J., Ed., IRL Press, Oxford, 1987, 71.

9. Robertson, E. J., Bradley, A., Kuehn, M., and Evans, M., *Nature*, 323, 445, 1986.

10. Chen, C. and Okayama, H., *Mol. Cell. Biol.*, 7, 2745, 1987.

11. Lovell-Badge, R. H., in *Teratocarcinomas and Embryonic Stem Cells — A Practical Approach*, Robertson, E. J., Ed., IRL Press, Oxford, 1987, 153.

12. Kim, H. S. and Smithies, O., *Nucleic Acids Res.*, 16, 8887, 1988.

13. McPherson, M. J., Ed., *PCR: A Practical Approach*, IRL Press, Oxford, 1991.

14. McPherson, M. J., Hames, B. D., and Taylor, G. R., Eds., *PCR2: A Practical Approach*, IRL Press, Oxford, 1995.

15. Edwards, M. C. and Gibbs, R. A., *PCR Methods Applic.*, 3, S65, 1994.

16. Sambrook, J., Fritsch, E. F., and Maniatis, T., *Molecular Cloning — A Laboratory Manual*, 2nd ed., Cold Spring Harbor Laboratory Press, Cold Spring Harbor, New York, 1989.

17. Wiles, M. V., in *Methods in Enzymology 225: Guide to Techniques in Mouse Development*, Wassarman, P. M. and DePamphilis, M. L., Eds., Academic Press, San Diego, 1993, 900.

18. Karasuyama, H. and Melchers, F., *Eur. J. Immunol.*, 18, 97, 1988.

19. Nakano, T., Kodama, H., and Honjo, T., *Science*, 265, 1098, 1994.

20. Potocnik, A. J., Nielsen, P. J., and Eichmann, K., *EMBO J.*, 13, 5274, 1994.

21. Miltenyi, S., Muller, W., Weichel, W., and Radbruch, A., *Cytometry*, 11, 231, 1990.

22. Austyn, J. M. and Gordon, S., *Eur. J. Immunol.*, 11 805, 1981.

23. Chomczynski, P. and Sacchi, N., *Anal. Biochem.*, 162, 156, 1987.

24. Golay, J., Passerini, F., and Introna, M., *PCR Methods Applic.*, 1, 144, 1991.

25. Souazé, F., Ntodouthome, A., Tran, C. Y., Rostene, W., and Forgez, P., *BioTechniques*, 21, 280, 1996.

26. Celi, F. S., Zenilman, M. E., and Shuldiner, A. R., *Nucleic Acids Res.*, 21, 1047, 1993.

Chapter

Defining the Function of Related Genes Using Null Mutant Mice

Vesa Kaartinen, Leena Haataja,
Nora Heisterkamp, and John Groffen

Contents

I.   Introduction

To define the biological function of closely related genes using traditional *in vitro* methods of biochemistry and/or cell biology has been a challenging task. Recently this situation has improved significantly due to the development of techniques which allow the manipulation of murine genes *in vivo*.[1] In order to assess the biological role of transforming growth factor-β3 (TGF-β3), a prototypic member of the TGF-β superfamily,[2-4] we generated a null mutant TGF-β3-deficient mouse strain.[5]

All three closely related, evolutionary-conserved mammalian TGF-βs (β1, β2, and β3) are encoded by genes located on different chromosomes.[3] They are synthesized as approximately 400 amino acid precursors from which the carboxy terminal 112 amino acid monomer is cleaved. Biologically active, mature TGF-βs are formed by dimerization of monomers. TGF-βs are secreted from cells as latent complexes, in which mature TGF-β homodimers are bound via noncovalent interaction to dimers of the glycosylated *N*-terminal precursor regions. These multi-functional molecules are thought to be involved in many biological processes, such as embryogenesis, wound healing, cardiac protection, hematopoiesis, and immune and inflammatory responses.[3,4] TGF-βs mediate a wide range of cellular activities including adhesion, proliferation, differentiation, extracellular matrix deposition, and transformation *in vitro*.[3,4,6] The biochemical activity of the mature form of the various isoforms *in vitro* is often qualitatively similar, but differs quantitatively, depending on the cell type used.[3,6,7] Distinct *in vitro* activities, however, have also been reported.[3] TGF-βs signal through two interacting TGF-β receptors, TGF-β RI and TGF-β RII, which are members of a rapidly expanding transmembrane Ser/Thr kinase family.[8]

Studies on expression levels during embryogenesis, regeneration, and pathological processes have strongly suggested that TGF-βs isoforms have biologically distinct functions.[3,9] Their characteristic spatio-temporal expression patterns during embryonic development[10-12] suggest discrete roles for each of the three TGF-β isoforms. Murine TGF-β3 transcripts are virtually undetectable during early embryogenesis before day 10 postcoitum (E10), but gradually increase from days E10.5 to E17.5. The medial edge epithelial cells of developing palatal shelves also show a very strong and specific expression of TGF-β3 transcripts at E13.5 through E14.5.[13,14] TGF-β3 mRNA is found at highest concentrations in these epithelial cells, at this period in embryogenesis.[13] In addition, clearly detectable TGF-β3 expression has been found in lung, cardiac, and skeletal muscle; chondrocytes in the ribs and vertebra; osteocytes; and intervertebral discs, spinal cord, brain, choroid plexus, and cochlear epithelium in the developing mouse.[10-12] In adult mouse tissues, TGF-β3 is highly expressed in the lung (particularly in airway epithelium), heart, brain, testis, and placenta, but little or no gene product is found in liver, spleen, and kidney.[6] A comparison of mRNA and protein expression patterns suggests that TGF-β3 acts through both autocrine and paracrine mechanisms.[10] Based on their accurately timed expression patterns *in utero,* a coordinated role for the respective TGF-β gene products in epithelio-mesenchymal interactions was suggested.[12]

To study the function of TGF-β during mouse development, we used gene targeting to introduce a null mutation in the mouse germ line.[5] Homozygous TGF-β3-deficient pups die shortly after birth and suffer from bilateral cleft of the secondary palate. The effects of its bi-allelic inactivation demonstrate an important role for TGF-β3 in normal palatal fusion. Moreover, with this TGF-β3 –/– mutation, an animal model is created for one of the most common disfiguring birth defects. This study clearly demonstrates that the knockout studies can be very helpful to define the specific biological function of one member of a family of highly related proteins.

II. Methods

A. Generation of Null Mutant Mice

1. General

Development of techniques allowing *in vitro* manipulation of genes in murine embryonic stem (ES) cells and subsequent generation of chimeras capable of transmitting the mutated allele to offspring has been one of the most important scientific achievements during the past 15 years in the field of biomedical research. Knockout mice have become irreplaceable tools for scientists studying functions of genes and gene products during biological processes such as embryogenesis and malignant transformation, and for the generation of animal models for human genetic diseases. Gene-targeting studies are particularly important for those applications in which traditional biochemical or transgenic approaches are often unable to give definitive answers (for instance in studies on the function of closely related genes).

The steps required for generation of null mutant mice include (1) preparation of a targeting vector, (2) transfection of the vector into pluripotent embryonic stem cells, (3) selection to enrich for the desired targeting events, (4) isolation and expansion of individual clones, (5) identification of the correctly targeted clones, (6) injection of the targeted embryonic stem cells into blastocysts, (7) crossing of chimeras to generate heterozygote mice, and (8) crossing of heterozygotes to generate null mutant mice (Figure 10.1).[15]

2. Targeting vectors and general strategies

Two different types of targeting vectors have been used: insertional vectors and (more commonly) replacement vectors.[16] The most frequently used positive and negative selection markers are the neomycin resistance (*neo*[r]) and thymidine kinase (*tk*) genes, respectively. Insertion of the positive selection marker into the coding exon of a gene disrupts the open reading frame, abolishing the function of the gene. In the presence of geneticin (G418) and ganciclovir (or 1[1-2-deoxy-2-fluoro-βD-arabinofurasyl]-5-iodouracil), only the cells which have retained the *neo*[r] gene but

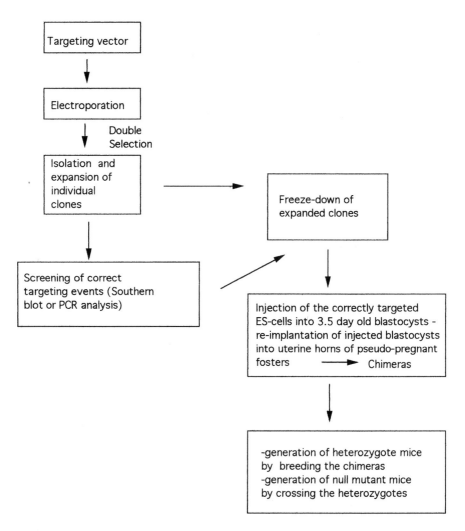

FIGURE 10.1
Strategy to generate targeted ES cell lines and null mutant mice.

have lost the negative selection marker, *tk,* will survive. Several improvements have been introduced during the past 5 years to enhance the targeting efficiency. Probably the most significant advancement is the use of isogenic targeting vectors, i.e., the targeting vector and ES cell line used will have an identical genetic background.[17] Using isogenic targeting vectors and double selection, the number of correctly targeted clones will often range from 1 to 30%. Recently, new techniques have been developed which allow (1) inactivation of genes in specific tissue, or (2) switching genes on and off at will, using systems such as tetracycline-regulated inducible gene expression or Cre-loxP-mediated gene targeting.[18-20]

3.   Embryonic stem cell cultures

Technically, the most demanding aspect of the generation of knockout mice is the maintenance and manipulation of the ES cells; they have to be kept (1) in an undifferentiated state, (2) free from possible tissue culture contaminants such as mycoplasma, and (3) free from karyotypic abnormalities. Failure to adhere strictly to any of these requirements will most likely prevent the generation of germ line chimeras. Most of the ES cell lines used nowadays have been derived from male mice of different substrains of 129Sv.[21] Long-term ES cell cultures can be grown (1) on mitotically inactivated feeder cells (STO, EMFI), (2) in the presence of buffalo/rat liver cell-conditioned medium, or (3) in the presence of recombinant leukemia inhibitory factor (LIF) in order to maintain their pluripotency. The combination of STO-feeder cells and recombinant LIF has worked very well in our hands.

B.   Strategy To Disrupt the TGF-β3 Gene

1.   Targeting vector to disrupt the TGF-β3 gene

A probe from the mature region of the human TGF-β3 cDNA was used to isolate a set of overlapping lambda clones from a partial *Mbo*I B6CBA/F1 genomic library. The TGF-β3 locus was mapped by restriction enzyme analysis. A 12-kb *Sal*I fragment containing exons 4, 5, 6, and 7 was partially digested with *Bsu36*I, and the pGK-neo-poly(A) cassette encoding the G418 resistance gene under control of the phosphoglycerate kinase promoter was inserted into the *Bsu36*I site of exon 6. This *Sal*I fragment was inserted into the *Xho*I site located in a thymidine kinase cassette, encoding thymidine kinase under the control of the herpes simplex virus promoter (Figure 10.2)

2.   Transfection and selection of embryonic stem cells

The R1 embryonic stem cell line (generously provided by Dr. A. Nagy) was cultured on a monolayer of STO feeder cells mitotically inactivated by mitomycin C in the presence of recombinant LIF (500 U/ml). Cells were trypsinized, resuspended at a concentration of 1×10^7 per ml in phosphate buffered saline (PBS), and electroporated (230 V, 500 μF) at room temperature with 30 μg of construct DNA linearized with *Not*I. Immediately after transfection, cells were plated and allowed to recover for 24 hr. Selection was carried out with G418 (150 μg/ml) and ganciclovir (2 μ*M*).[5] Double-drug-resistant colonies were picked 8 to 11 days after electroporation. Each individual clone was expanded and frozen down. An aliquot of each clone was analyzed for homologous recombination by Southern blot analysis using *Eco*RI digestion and an external probe-a (Figure 10.2). The correct targeting was verified using *Bsu36*I digestion and hybridization with probe-a. The frequency of correct targeting at the TGF-β3 genomic locus was 1 in 65 double-resistant clones. The correctly targeted clone 56.1 was analyzed by Southern blot in order to verify a single copy insertion and a loss of the *tk* gene using *neo* and *tk* probes, respectively.

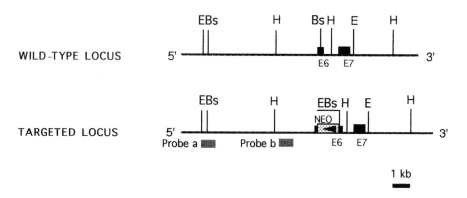

FIGURE 10.2
Schematic representation of the mouse TGF-β3 locus and its inactivation by homologous recombination. Exons 6 and 7 are indicated as black boxes. External probe-a and internal probe-b are shown as shaded boxes. Restriction enzymes: E, *Eco*RI; Bs, *Bsu*36I; H, *Hind*III.

Moreover, the possibility of gross numerical karyotypic abnormalities such as aneuploidy was excluded by counting metaphase chromosomes.

3. Injection of the correctly targeted embryonic stem cells into blastocysts and generation of chimeras

Clone 56.1, which contained a correctly targeted TGF-β3 gene, was used for blastocyst injections. ES cells (12 to 15) were injected into the blastocoelic cavity of C57BL/6 blastocysts at day 3.5 postcoitum. Injected blastocysts were transferred to uterine horns of B6/CBA (Fl) pseudopregnant females (mated with vasectomized Swiss-Webster males). Chimeras were identified by agouti coat color, and male chimeras were mated with Black Swiss females. Offspring with agouti coat color were genotyped using Southern blot analysis.

4. Southern blot assays

The genotype assays of mice were performed using *Hind*III and hybridization with an internal probe-b (Figure 10.2)

5. Expression studies

Expression of TGF-β3 was studied using a reverse transcription-polymerase chain reaction (RT-PCR) method. Briefly, total RNA was isolated from embryos. cDNAs were synthesized using a reverse primer (nucleotides 1659 to 1680 of the TGF-β3 cDNA[22]) and PCR performed using the reverse primer and a forward primer corresponding to nucleotides 1233 to 1254. Quality of RNAs was evaluated using b-actin as an internal standard.

6.   Histological studies

Tissues were fixed in 10% buffered formalin and stained with hematoxylin and eosin. Alizarin red and Alcian blue staining was performed as described.[23]

III.   Results and Discussion

A targeting vector containing a 12-kb genomic segment was constructed in which the positive neomycin resistance selection marker was inserted into exon 6 (Figure 10.2).[5] This insertion was expected to destroy almost the entire mature region of TGF-β3. The construct was electroporated into R1 embryonic stem cells and correct targeting was obtained. A correctly targeted clone was injected into C57BL/6 blastocysts, and highly chimeric mice were generated. Heterozygotes were obtained by breeding the chimeric male mice with Black Swiss females, and subsequently null mutant TGF-β3 –/– mice were obtained by cross-breeding the heterozygotes. Genotype analysis of offspring immediately after birth demonstrated the expected Mendelian ratio of wild-type, heterozygous, and homozygous offspring (28% +/+, 47% +/–, 25% –/– pups; n = 220). This indicated little or no prenatal mortality; however, TGF-β3 pups did not suckle and died within 11 to 20 hr after birth. RT-PCR analysis verified the absence of TGF-β3 transcripts in null mutants. Despite the comparable birthweight of TGF-β3 –/– and +/+ animals, the condition of the null mutant pups deteriorated rapidly and eventually they became cyanotic and appeared to be gasping for air.

TGF-β3 is strongly expressed in developing bones; therefore, we first studied the skeletal phenotype in detail. Surprisingly, these studies did not reveal significant skeletal abnormalities in null mutant fetuses or newborns (Figure 10.3). Instead, all null mutant pups suffered from cleft palate (Figure 10.4) The palatal cleft proceeded into the anterior part of the palate in half of the homozygotes (34/67), exposing the nasal cavities, whereas in 33 out of 67 homozygotes the anterior segment appeared to be fused. In many cases, fusion of the primary palate to the secondary palate was also defective. The palatal phenotype was not dependent on the genetic background since this was similar in both C57BL/6 and Black Swiss strains (3 backcrosses). Palatogenesis occurs relatively late in mouse development (E13 to E14.5) and is governed by epithelial-mesenchymal interactions.[24,25] Vertical growth and elevation of the wild-type and null-mutant palatal shelves were comparable in terms of size and shape as shown by serial coronal sections along the entire length of the palate. Despite clear contact and adherence of null mutant palatal shelves at E14.5, they failed to fuse. Oral and nasal aspects of null mutant epithelia had differentiated to pseudo-ciliated columnar cells and stratified squamous cells, respectively, comparable to that of wild-type. Continuous mesenchyme between the two sides of the wild-type palate was detected at E16.5, whereas the palatal fusion had not occured in null mutant specimen, but there was a cleft of the secondary palate.

During palatogenesis, expression of TGF-β3 is limited to the medial edge epithelial cells of prefusion palatal shelves.[13,14] After the midline epithelial seam is formed, TGF-β3 expression ceases rapidly and the medial edge epithelial cells

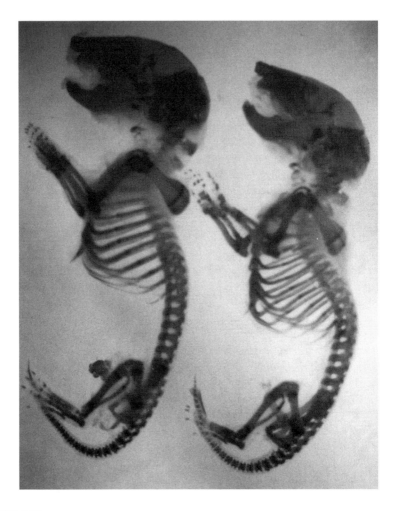

FIGURE 10.3
Skeletal phenotype of the TGF-β3 null mutant fetus (E18.5) (right) is indistinguishable from that of its wild-type littermate (left).

transdifferentiate to mesenchymal cells.[26-28] Despite the normal growth and elevation of TGF-β3 null mutant palatal shelves, they fail to fuse, although they are clearly adherent and in contact with each other. This clearly demonstrates an essential and unique role of TGF-β3 in palatal fusion.

The mouse model generated is exceptional in that TGF-β3 null mutants consistently exhibit a palatal defect, without any other detectable craniofacial abnormalities as compared to other null mutant mouse models with cleft palate syndrome.[29-33] However, the proposed role of follistatin in the regulation of TGF-βs is supported because the palatal phenotype seen in the TGF-βs –/– mice resembles some features of follistatin –/– mice.[33] In mice, the teratogenic compound 2,3,7,8-tetrachlorodibenzo-

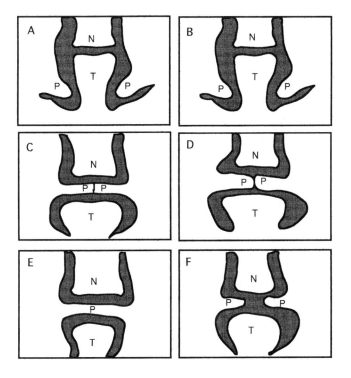

FIGURE 10.4

Schematic representation of frontal sections through the heads of days E13.5, E14.5, and E16.5 wild-type (A, C, E) and null mutant (B, D, F) embryos. At day E13.5, the wild-type (**A**) and null mutant (**B**) palatal shelves (P) grow vertically along the sides of the tongue (T). At day E14.5, wild-type shelves are fusing, with the midline seam still clearly visible (**C**), whereas null mutant shelves are adherent but show no signs of epithelial fusion (**D**). Two days later (E16.5), wild-type shelves are properly fused (**E**), while null mutant shelves show a cleft of the secondary palate (**F**). N = nasal septum.

p-dioxin (TCDD) induces cleft palate,[34] which closely resembles the palatal defect of TGF-β3 –/– mice, i.e., fully grown shelves fail to fuse. This suggests that the aryl hydrocarbon receptor (AHR) affects or is an upstream regulator of TGF-β3, since the action of TCDD is mediated through the AHR receptor.[35]

Targeted disruption of a closely related TGF-β gene, TGF-β1, causes either a severe multifocal inflammatory disease[36,37] at the age of 2 to 3 weeks (50% of the offspring) or a failure of vasculogenesis and hematopoiesis around E10 (50% of the offspring).[38] Although TGF-β1 has been shown to cross the placental barrier, the relative amounts of TGF-β1 in the null mutant offspring compared to wild-type fetuses are unknown.[39] Due to the significant biochemical similarity between TGF-β1 and TGF-β3, the latter will likely cross the placental barrier as well. Since TGF-β3-null mutants die shortly after birth, a potential role of maternal rescue on the null-mutant phenotype remains unknown. However, the strong and highly specific expression of TGF-β3 in the medial edge epithelium of the palatal shelves before and during the palatal fusion and the fact that all TGF-β3-null mutants suffer from cleft palate, indicate that a high local concentration of TGF-β3 gene product is needed for successful palatal fusion. This, in turn, would suggest that a mode of action is autocrine or local paracrine, rendering the effect of maternal rescue insignificant during this process.

Before the advent of knockout studies, expression studies were used to evaluate the role of each individual TGF-β isoform during development.[10-12] The fact that expression patterns are temporally strictly controlled, and spatially both overlapping and distinct, was used as an argument to suggest that TGF-βs play a role both in differentiation and morphogenesis during mouse development; however, expression studies were unable to determine the exact biological role(s) of each isoform and to evaluate processes in which TGF-βs are able to compensate for each other's functions. Recent gene-targeting studies have clearly provided valuable new data on the biological function of both TGF-β1[36-38] and TGF-β3[5,40] and have pinpointed a role for each of these closely related molecules in distinct developmental processes. TGF-β1 was shown to be essential for a normal inflammatory response, as well as for normal formation of blood vessels and blood cells, whereas TGF-β3 is the key component in the complex process of palatal fusion.

The development of the murine palate is an excellent example of the power of gene-targeting studies in defining the role of closely related genes in biological processes, since all TGF-βs are expressed in different subcompartments of prefusion palatal shelves;[13,14] however, the loss of TGF-β1 does not lead to palatal clefting, whereas our gene-targeting study has clearly shown the importance of TGF-β3 in this process. A different example of a process in which TGF-βs most likely are able to compensate for each other's function is the development of the skeletal system. All TGF-βs are strongly expressed in developing bones, but neither the TGF-β1 nor TGF-β3 null mutants display clear skeletal abnormalities.

Cleft palate is one of the most common human birth defects.[24] Similarities between the human overt cleft palate without associated craniofacial abnormalities and the TGF-β3-null mutant palatal phenotype implicate TGF-β3 in the pathogenesis of human cleft palate. Our study shows a mechanistic defect in the epithelial-mesenchymal interaction in TGF-β3 –/– null mutant mice and provides an animal model for a human cleft palate syndrome. Most importantly, phenotypic characterization of the TGF-β3 –/– mice clearly demonstrates that gene targeting by homologous recombination is a powerful tool to define *in vivo* functions of closely related genes.

References

1. Capecchi, M., *Science,* 244, 1288, 1989.

2. Kingsley, D. M., *Genes Dev.,* 8,133,1994.

3. Massague, J., *Ann. Rev. Cell Biol.,* 6, 597, 1990.

4. Roberts, A. B. and Sporn, M. B., in *Peptide Growth Factors and Their Receptors, Handbook of Expermental Pharmocology,* Sporn, M. B. and Roberts, A. B., Eds., Springer-Verlag, Heidelberg, 1990, 419.

5. Kaartinen, V., Voncken, J. W., Shuler, C., Warburton, D., Bu, D., Heisterkamp, N., and Groffen, J., *Nature Genet.,* 11, 415, 1995.

6. Miller, D. A., Pelton, R. W., Derynck, R., and Moses, H., *Ann. N.Y. Acad. Sci.,* 593, 208, 1990.

7. Graycar, J. L., Miller, D. A., Arrick, B. A., Lyons, R. M., Moses, H. L., and Derynck, R., *Mol. Endocrinol.,* 3, 1977, 1989.

8. Wrana, J. L., Attisano, L., Wiesner, R., Ventura, F., and Massague, J., *Nature,* 370, 341, 1994.

9. Roberts, A. B. and Sporn, M. B., *Molec. Reprod. Dev.,* 32, 91, 1992.

10. Schmid, P., Cox, D., Bilbe, G., Maier, R., and McMaster, G. K., *Development,* 111, 117, 1991.

11. Millan, F. A., Denhez, F., Kondaiah, P., and Akhurst, R. J., *Development,* 111, 131, 1991.

12. Pelton, R. W., Dickinson, M. E., Moses, H. L., and Hogan, B. L. M., *Development,* 110, 609, 1990.

13. Fitzpatrick, D. R., Denhez, F., Kondaiah, P., and Akhurst, R. J., *Development,* 109, 585, 1990.

14. Pelton, R. W., Hogan, B. L. M., Miller, D. A., and Moses, H. L., *Dev. Biol.,* 141, 456, 1990.

15. Capecchi, M. R., *Trends Genet.,* 5, 70, 1989.

16. Hasty, P. and Bradley, A., in *Gene Targeting — A Practical Approach,* Joyner, A. L., Ed., Oxford University Press, Oxford, 1993, 1.

17. te Riele, H., Maandag, E. R., and Berns, A., *Proc. Natl. Acad. Sci. USA,* 89, 5128, 1992.

18. Gu, H., Zou, Y.-R., and Rajewsky, K., *Cell,* 73, 1155, 1993.

19. Kühn, R., Schwenk, F., Aguet, M., and Rajewsky, K., *Science,* 269, 1427, 1995.

20. Wurth, P. A., St. Onge, L., Boger, H., Gruss, P., Gossen, M., Kistner, A., Bujard, H., and Henninghausen, L., *Proc. Natl. Acad. Sci. USA,* 91, 9302, 1994.

21. Papaioannou, V. P. and Johnson, R, in *Gene Targeting — A Practical Approach,* Joyner, A. L., Ed., Oxford University Press, Oxford, 1993, 107.

22. Miller, D. A., Lee, A., Matusui, Y., Chen, E. Y., Moses, H. L., and Derynck, R., *Mol. Endocrinol.,* 3, 1926, 1989.

23. McLeod, J. M., *Teratology,* 22, 299, 1980.

24. Ferguson, M. W. J., *Development,* 103s, 41, 1988.

25. Sharpe, P. M. and Ferguson, M. W. J., *J. Cell. Sci. Suppl.,* 10, 195, 1988.

26. Fitchett, J.E. and Hay, E. D., *Dev. Biol.,* 131, 455, 1989.

27. Shuler, C. F., Guo, Y., Majumder, A., and Luo, R., *Int. J. Dev. Biol.,* 36, 463, 1991.

28. Shuler, C. F., Halpern, D. E., Guo, Y., and Sank, A. C., *Dev. Biol.*, 154, 318, 1992.

29. Rijli, F. M., Mark, M., Lakkaraju, S., Dierich, A., Dolle, P., and Chambon, P., *Cell*, 75, 1333, 1993.

30. Gendron-Maquire, M., Mallo, M., Zhang, M., and Gridley, T., *Cell*, 75, 1317, 1993.

31. Satokata, I. and Maas, R., *Nature Genet.*, 6, 348, 1994.

32. Matzuk, M. M., Kumar, T. R., and Bradley, A., *Nature*, 374, 356, 1995.

33. Matzuk, M. M., Lu, N., Vogel, H., Sellheyer, K., Roop, D. R., and Bradley, A., *Nature*, 374, 360, 1995.

34. Couture, L. A., Abbott, B. D., and Birnbaum, L. S., *Teratology*, 42, 619, 1990.

35. Abbott, B. D., Perdew, G. H., and Birnbaum, L. S., *Toxicol. Appl. Pharmacol.*, 126, 16, 1994.

36. Shull, M. M., Ormsby, I., Kier, A. B., Pawlowski, S., Diebold, R.J., Yin, M., Allen, R., Sidman, C., Proetzel, G., Calvin, D., and Doetschman, T., *Nature*, 359, 639, 1992.

37. Kulkarni, A. B., Huh, C. G., Becker, D., Geiser, A., Lyght, M., Flanders, K. C., Roberts, A. B., Sporn, M. B., Ward, J. M., and Karlsson, S., *Proc. Natl. Acad. Sci. USA*, 90, 770, 1993.

38. Dickson, M. C., Martin, J. S., Cousins, F. M., Kulkarni, A. B., Karlsson, R. A., and Akhurst, R. J., *Development*, 121, 1845, 1995.

39. Letterio, J.J., Geiser, A. G., Kulkarni, A. B., Roche, N. S., Sporn, M.B., and Roberts, A.B., *Science*, 264, 1936, 1994.

40. Proetzel, G., Pawlowski, S. A., Wiles, M. V., Yin, M., Boivin, G. P., Howles, P. N., Ding, J., Ferguson, M. W., and Doetschman, T., *Nature Genet.*, 11, 409, 1995.

Chapter

11

Genetic Analysis of Mouse Models for Rheumatoid Arthritis

*Rikard Holmdahl, Stefan Carlsén,
Anna Mikulowska, Mikael Vestberg,
Ulrica Brunsberg, Ann-Sofie Hansson,
Mats Sundvall, Liselotte Jansson,
and Ulf Pettersson*

Contents

215

I.   Introduction

Rheumatoid arthritis (RA) is an endemic disease affecting a large part of the world population. In spite of intense research, the etiology and pathogenesis are poorly known; consequently, the disease can be neither prevented nor cured. Current treatments are largely directed to symptoms and do not change the ultimate disease course. The disease is characterized by a tissue-specific inflammatory attack directed to peripheral joints. The inflammation develops in typical cases into a relapsing pattern with recurrent attacks in the same or other joints and with secondary systemic inflammatory lesions. This results in a chronic disease course which often progressively disables the individual.

A.   Genetic Analyses of Rheumatoid Arthritis

Rheumatoid arthritis affects approximately 1% of the Caucasian population and afflicts all other human populations to a variable degree. Studies of monozygotic twins indicate a relatively weak but significant genetic influence with a concordance rate between 10 and 30%. It has been clearly documented that RA is associated with the major histocompatibility complex (MHC). MHC encodes a series of cell surface receptors of crucial importance for the activity of the immune system. The MHC class I and class II receptors bind antigenic peptides presented to T lymphocytes which enable them to give rise to an antigen-specific immune response. The MHC class I and class II genes are highly polymorphic, explaining the fact that the specific immune responsiveness varies between individuals. This has supported the argument that RA could be an autoimmune inflammatory disease. Hypothetically, the immune system recognizes peptides derived from the specific tissues and responds to them as if they were of foreign, or infectious, origin. A T-cell-directed inflammatory response to the cartilaginous joints ensues. The responsible autoantigenic peptides, however, have not been isolated. It is also unclear why the immune system is not tolerant to tissue-specific antigens. Development of classical RA is associated with the MHC class II molecule DR4. DR4 alleles, however, are common in the general population, and the vast majority of carriers do not develop RA. Thus, a search for other genes as well as environmental factors is needed to understand the etiology and pathogenesis of RA.

The search for predisposing genes for development of RA is hampered by several factors.

- First, RA is defined by a number of phenotypic criteria; therefore, it is most likely that RA comprises several distinct diseases sharing common symptoms.
- Second, the human population is highly heterogeneous, making the search for specific genotypes difficult.
- Third, uncontrolled and unknown environmental factors are likely to play an important role.
- Finally, the disease is controlled by many genes, most likely with varying penetrance; moreover, different genes may be operating in different affected families (genetic heterogeneity).

Not surprisingly, it has been difficult to define clearly associated loci by gene screening of large numbers of sibling pairs, whereas more restricted analyses of family groups show significant linkage of certain loci possibly of importance for the disease.

B.   Animal Models

A shortcut in the genetic analysis of RA is provided by the animal models — both for direct comparisons with the human diseases and, maybe more importantly, for the understanding of the mechanisms in the pathogenesis. For studies of RA, several animal models have been used. One common model is to induce disease with cartilage-specific type II collagen (CII)[1] — the so-called collagen-induced arthritis (CIA). Induction of CIA with heterologous CII (of non-mouse origin) in the mouse is a well-characterized model, but the disease does not develop into chronicity. In contrast, induction of CIA with autologous CII in the rat leads to a chronic disease course and may therefore provide a better model for RA but one that is more difficult to analyze genetically. We will here describe the induction and genetic analysis of CIA in the mouse. We will also give an example of how to induce and follow the disease in a highly susceptible mouse strain.

TABLE 11.1
CIA in Common Inbred Strains and in Mice
Carrying Selected Mutations

Strain	MHC[a]	Mutation	CIA[b]	SIA[c]	Anti-CII Response
DBA/1	q		+++++	+++++	++++
B10.Q	q		++	–	+++
C3H.Q	q		++++	–	++++
SWR	q	C5 –/–, deficiency of complement 5	–	nd	+++++
NFR/N	q		++++	–	++++
B10.RIII	r		+++++	–	++++
RIIIS/J	r		–	nd	++++
DBA/1-yaa	q	yaa; Y chr autoimmune accelerating gene, promotes lupus	+	+	++
DBA/1-xid	q	xid; bruton tyrosin kinase deficiency affects B-cell signaling	–	nd	+

[a]   MHC: major histocompatibility complex.

[b]   CIA: collagen-induced arthritis.

[c]   SIA: stress-induced arthritis.

II.   Collagen-Induced Arthritis

A.   Genetic Influences

There is a large variability in the susceptibility to CIA among inbred mouse strains. A major factor is the MHC dependence; in fact, a class II gene has been defined (see below). There is also an influence by many other so far unknown genes outside MHC. A number of spontaneously occurring gene mutations in inbred strains have given some clues to the pathogenesis of CIA. A functional gene for complement (C5) is important[2] but not absolutely necessary[3] for development of CIA, emphasizing the role of antibodies. In addition, gene mutations affecting B-cell function have been shown to lead to resistance to CIA.[4,5] The susceptibility to CIA in a selected number of inbred strains is given in Table 11.1.

B.   Environmental Influence

Susceptibility to CIA is also strongly dependent on environmental factors (Table 11.2).

TABLE 11.2
Some Environmental Effects
on Mouse CIA and SIA

Environmental Effect	Effect on CIA	Effect on SIA
Inter-male stress	+	+
Pregnancy	–	Not applicable
Estrogen	–	Not done
Testosterone	+	+
Darkness (melatonin)	+	+

Note: + = exaggerated arthritis; – = ameliorated arthritis; CIA = collagen-induced arthritis; SIA = stress-induced arthritis.

1. Light

Mice kept in darkness are more susceptible to CIA and develop a higher anti-CII antibody response.[6] This effect is most likely dependent on melatonin induction.[7,8]

2. Stress

The stress caused by inter-male aggressiveness as a consequence of grouping male mice may dramatically enhance the susceptibility to arthritis. Thus, the number of DBA/1 male mice housed together in a cage matters for CIA development, since grouped mice show an enhanced susceptibility to CIA induction compared to mice housed alone. The importance of such influence differs among various mouse strains (i.e., different genetic backgrounds). Induction of inter-male aggressiveness will lead to spontaneous development of arthritis in the DBA/1 strain.[9] This stress-induced arthritis (SIA) develops only if the males from different litters are grouped more than two in each cage. Castrated mice do not develop arthritis, whereas testosterone treatment of castrated male DBA/1 mice leads to susceptibility. When castrated, the males changed behavior and stopped fighting, which started again during treatment with testosterone. This clearly demonstrates a connection between behavior and arthritis development in DBA/1 males. DBA/1 females are resistant to arthritis induced by stress.

3. Sex

As in RA there is a sex difference in CIA susceptibility in mice, although it is reversed as compared with the human disease. Male mice are more susceptible to disease development than females due to the influence of estrogen.[10] Castration of females renders them as susceptible as male mice. Females are less susceptible during pregnancy but develop a disease relapse postpartum — an effect attributed to estrogen levels.[11,12] The immunosuppression of castrated females after treatment with estrogen is both dependent on the dose of estrogen and on the genetic background.[13]

4. Age

Only adult mice are susceptible to arthritis. DBA/1 mice need to be at least 6 weeks of age. After puberty the age does not have a dramatic influence, although male DBA/1 mice tend to be more susceptible with age,[14] an effect which could be related to the behavior as discussed above.

5. Infections

The role of infections in the environment is not clear for development of CIA in mice; however, CIA experiments tend to be more variable in facilities with an uncontrolled environment. We therefore recommend that all experiments be performed in specific pathogen-free (SPF) controlled facilities in which the infectious load is defined and stable.

C. Performing a Mouse Collagen-Induced Arthritis Experiment

1. Induction

Due to the sensitivity to behavioral effects on the susceptibility to CIA, the experiments should be planned a long time in advance. At the time of weaning or when mice obtained from a breeding facility arrive, create cages with the appropriate number of mice in each cage. A susceptible strain is a cross between the DBA/1 and B10.Q strains, (B10.Q×DBA/1)F1, which can be purchased from Bomholtgård (Denmark, fax +45/86841699). The mice should rest at least 2 weeks before starting the experiment.

Mark all mice. The best method is to toemark the mice when they are newborn. This is a secure and, for the mice, the mildest way for marking. The distal phalang of a toe is taken in a system that allows distinction of an almost unlimited number of mice. Adult mice can be earmarked or marked with a chip. Collect material for DNA-preparation either from the toe phalang or from the tip of the tail. The mice should be at least 7 weeks old when immunized and should be kept in equal numbers in each cage, normally 3 to 10 mice in a standard type III cage. Be aware that mixing male mice at the time of immunization (of certain strains, such as DBA/1) from different litters will incite fighting, which affects the disease outcome. It is important to mix them randomly within the cages. The reason for doing so is that arthritis occurrence is usually not uniformly distributed in cages; if arthritis occurs in one cage, the cage-mates tend to get arthritis more often. This could possibly be due to stress effects and will give false results if mice kept in different cages are compared. The mice should be scored before immunization, in accordance with the selected scoring protocol to be used (see Appendix A).

For the immunization, it is important to use pure native CII. Advice for the collagen preparation procedure is given in Appendix B. Arthritogenic CII derived from various species are commercially available from Chondrex (Seattle, WA; fax

206-882-3094). In our protocol, freeze-dried type II collagen (CII) dissolved in ice-cold $0.1\text{-}M$ acetic acid at a concentration of 2 mg/ml is thoroughly mixed with an equal volume of complete Freund's adjuvant (Difco; Detroit, MI) for a 1:1 ratio. By using a homogenizer or a syringe with a 18G × 2-in needle, a thick white emulsion is obtained. When all the water solution is dispersed in the oil, a drop of the emulsion holds together when placed on a water surface. Another way of testing the emulsion is by centrifugation (3000 rpm for 30 sec). If there are no separate phases of water, emulsion, and oil, it is ready. A glass syringe (1-ml), with a Luer-Lok and a 27G × 3/4-in needle, is used for immunization intradermally (i.d.). The Luer-Lok connection is important, since the emulsion is viscous and the needle is very thin in order to make i.d. injection possible.

The mouse is sedated with a light anesthetic. Brush the fur at the root of the tail upwards. Place the tip of the needle 3 to 4 mm down on the tail and enter the skin. Push the needle very gently and superficially in the skin about 10 mm along the back of the mouse. At the base of the tail, inject 100 µl of the emulsion, place thumb on the place of entry of the needle, and remove the needle. Keep the pressure on the tail with the thumb for about 5 sec to prevent the emulsion from leaking. If the emulsion is correctly injected i.d., there is a distinct white cushion seen in the skin. If the emulsion is injected subcutaneously, it will spread into a large area.

2.   Inspection

Inspection of mice for occurrence of arthritis should start at 14 days after immunization. Three to 7 weeks after CII immunization, the arthritis suddenly appears in one of the peripheral joints in a paw and usually will subsequently involve additional joints. Erythema and swelling are the first signs which within a few hours or a day develop into severe arthritis. Often one or two toe joints are first affected, after which the arthritis spreads to include finally the entire paw. The acute inflammation, as indicated by erythematous swelling, usually persists for about 2 to 3 weeks before starting to decline; subsequently, a healing process resulting in deformities and stiffness usually follows. In mice, the disease course is usually monophasic. The best way to document the clinical progression of the arthritis is by visual inspection using a scoring system (see Appendix A and Figure 11.1). If appropriately performed, this will give the most adequate reflection of the disease. It is, however, very important to follow strict rules for the scoring. The most important are that the scoring must be performed in a "blinded" manner and a fixed scoring protocol must be used allowing two different trained investigators to score the mice independently with the same result.

There are also other useful ways to measure arthritis. In our experience, the best parametric test is to measure thickness of the joints using a caliper or a micrometer screw. This requires some training to give reproducible results and is therefore also dependent on the investigator. In addition, it is not possible to include toe joints in the measurements, and the test does not distinguish between active arthritis and deformed joints; however, it may have an advantage as a measurement complementary to scoring.

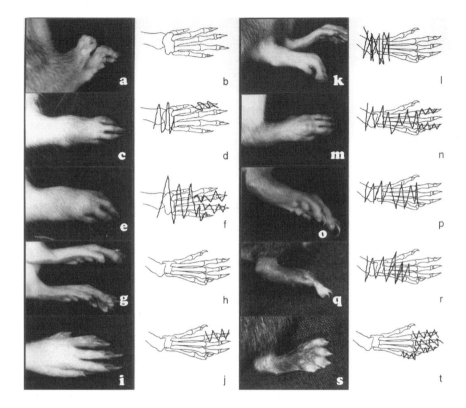

FIGURE 11.1

Examples of arthritic paws and corresponding line drawings depicting swelling and erythema. Scoring values for both the simple scoring system (S) and the extended scoring system (E) are given. The scoring is applicable for both mice and rats. (**a, b**) Front paws from normal DA rat: S = O, E = O. (**c, d**) Front paw from arthritic LEW rat; the ankle and one toe are inflamed: S = 4, E = 6 (5A [ankle] +1T [toe]). (**e, f**) Front paw from arthritic LEW rat; the ankle, the midpaw, and three toes are inflamed: S = 4, E = 13 (5A + 5M [midpaw] + 3T). (**g, h**) Hind paws from normal LEW rat: S = O, E = O. (**i, j**) Hind paw from arthritic LEW rat; one toe is inflamed: S = 1, E = 1. (**k, l**) Hind paw from arthritic DA rat; the ankle is inflamed and stiff: S = 3, E = 6 (5A + 1 [stiffness]). (**m, n**) Hind paw from arthritic LEW rat; the ankle midpaw, and two toes are inflamed: S = 4, E = 12 (5A + 5M + 2T). (**o, p**) Hind paw from arthritic LEW rat; the ankle and the midpaw are inflamed: S = 3, E = 10 (5A + 5M). (**q, r**) Hind paw from arthritic C3H mouse; the ankle and the midpaw are inflamed: S = 2, E = 10 (5A + 5M). (**s, t**) Hind paw from arthritic C3H mouse; the midpaw and four toes are inflamed: S = 2, E = 9 (5M + 4T).

Arthritis often appears between days 21 and 40 after immunization. Onset day is an important phenotype; thus, inspection should be performed at least 3 times per week starting on day 18. When the paws have been severely swollen for a week or two, gross deformities of the paws usually appear. As the swelling subsides, the entire paw shows deformity in the form of unnatural joint angles, stiffness, and a shriveled appearance.

TABLE 11.3
Blood Sampling from Mice

1. Place the mouse on the bench and pull the tail through a cage lid or anesthetize the mouse.

2. Cut a tail vein with a scalpel or bleed by puncture of the venous plexus behind the eye (retro-orbital puncture) using a Pasteur pipette or a capillary tube.

3. Collect not more than 200 μl blood in a tube. Shake the tube after bleeding.

4. Let the blood coagulate for 4 hr at room temperature or overnight at 4°C.

5. Centrifuge and collect the serum.

6. Store serum at −20°C.

3.   Bleeding

To follow the immune response to CII, the antibody titers to autologous CII are measured. The mice are bled at day 35 when the anti-CII titers are maximal and during a time when they are in the process of developing arthritis (see Table 11.3). The serum is quantitated for antibodies to CII in ELISA as described in detail elsewhere.[15] The B-cell response to CII is a valuable phenotype, since the disease is to a large extent dependent on pathogenic antibodies driving the inflammatory attack on the joints.

4.   Termination

The experiment can be terminated when no more mice develop arthritis and when all previously arthritic joints are inactive. This usually happens before day 65. Histological examination of the paws at this stage will reveal various stages of healing and if some of the mice still have an ongoing subclinical inflammation. If histology of active arthritis is selected as a phenotype, the experiment must be terminated before the healing has occurred. Scoring of both active arthritis and the healing process is explained in Appendix C and Figure 11.2. Screening of histopathology can, however, sometimes be misleading, depending on the selection of paws and how they are sectioned, and will therefore require participation of trained independent investigators.

III.   Genetic Mapping of Experimental Models

Genetic mapping in inbred line crosses is generally much simpler than genetic mapping in outbred populations. Because inbred parental lines are homozygous for all loci, the *F1* generations from a cross between two inbred parental lines are heterozygous for all loci, and these individuals are fully informative for linkage

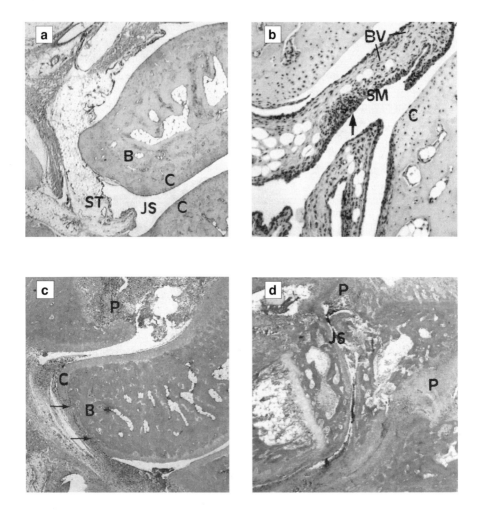

FIGURE 11.2

Photographs of histopathology of arthritic paws showing examples of inflammation and healing scores. The scoring is applicable for both mice and rats. (**a**) *Normal tissue:* A rat ankle with normal structures such as joint space (JS), cartilage (C), bone (B), and synovial tissue (ST) (Htx-e staining, original magnification 25×). (**b**) *Score 1:* A rat ankle with villus formation and a mild synovitis in the tissue. The hyperplastic synovial membrane (SM), increased numbers of blood vessels (BV), and small foci of inflammatory cells (arrow) are all characteristic markers of score 1. There are no bone or cartilage erosions (Htx-e stain, original magnification 100×). (**c**) *Score 2:* A rat ankle with obvious erosion of the cartilage and a smaller erosion of the talus bone (arrows). A moderate synovitis with pannus formation (P) is also seen but still with an undisrupted joint architecture (Htx-e staining, original magnification 25×). (**d**) *Score 3:* Metatarsal area in a rat paw with severe pannus, extensive erosions of bone and cartilage, and a disrupted joint architecture (Htx-e staining, original magnification 25×). (**e**) *Score I:* A rat ankle in early healing process with new formation of cartilage (CF) but no visible bone formation (Safranin-O stain, original magnification 100×). (**f**) *Score II:* A rat ankle with new bone formation (BF) and cartilage in tibia (Safranin-O stain, original magnification 25×). (**g**) *Score III:* Metatarsal area in late healing phase with new bone and cartilage formation and also with apparent ankylosis (A) (Safranin-O stain, original magnification 25×).

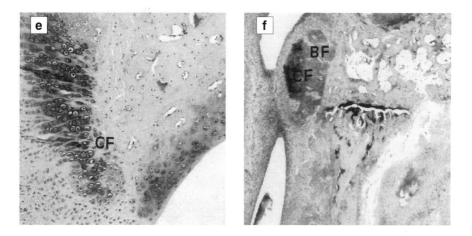

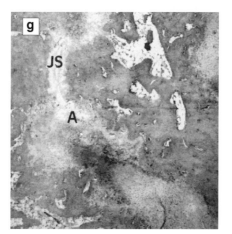

FIGURE 11.2 (continued)

mapping (to be informative for linkage mapping an individual has to be heterozygous both at the marker locus and at the trait locus). When genetic mapping is performed in outbred populations, only a subset of the individuals are informative for linkage and they are therefore less powerful for detecting linkage. In crosses that stem from inbred lines, there are only two alleles that segregate at each locus compared to the possibility that several alleles segregate at each locus in outbred populations. In inbred line crosses, the linkage phase is known, which is generally not the case in outbred populations, meaning that both phases have to be evaluated for each parent offspring pair. Altogether this means that mapping of inbred line crosses can be made simply by regressing the average of the trait values for all individuals onto the genotype.

The experiment crosses that are used most often are the *F2* backcross and the *F2* intercross. Both crosses start with *F1* individuals, produced by crossing two parental strains, *P1* (QQ genotype) and *P2* (qq genotype), drawn from inbred lines. All individuals in the *F1* generation will be heterozygous (Qq genotype). If two individuals from the *F1* generation are crossed, the result is an *F2* intercross population where the trait will segregate and the three different genotypes — QQ, Qq, and qq — are present in a 1:2:1 ratio. A backcross can be made by crossing an *F1* individual to either the *P1* or *P2* parent. The resulting genotypes in the two different crosses are QQ,Qq or qq. The *F2* intercross has an advantage over the *F2* backcross in that it allows the estimation of dominance.

Ideally, the trait is a quantitative measurement where the values are normally distributed. Often the measurements have a distribution that is not normal. Sometimes these measurements can be transformed (i.e., by taking the logarithm of the values) to make the distribution normal. If the trait is not continuous (i.e., score based on visual inspection of the individual) quantitative trait loci (QTL) mapping cannot be used, and other methods based on nonparametric statistics must be used. If a quantitative trait is available, the most straightforward way to detect it is to compare phenotypic means of different single-locus marker genotypes. A *t*-test can then be used to compare the means if only two marker genotypes are present, such as in a backcross (QQ,Qq). If three marker genotypes are present, such as in an *F2* intercross (QQ,Qq,qq), other methods have to be used such as ANOVA or regression analysis. A significant marker variance or a significant correlation coefficient, r^2, indicates the presence of a linked QTL. Usually the QTL effect is described as an additive or dominance effect of the allele derived from one of the parents. If an *F2* experimental cross has been made with inbred parental strains that are A/A and B/B where the B allele contributes to the phenotype, the measured phenotype of individual *i* is given by:

$$f_i = m + a \times B_i + d \times H_i + \varepsilon \qquad (11.1)$$

where *m* is the mean value of the component of the trait not controlled by this QTL, which is also the average phenotypic value of the A/A strain; *a* is the additive component of the QTL B-allele effect; B_i is the number of B alleles carried by individual number *i* (0, 1, or 2); *d* is the dominance component of the QTL B-allele effect; H_i is one if individual *i* is a heterozygote, and zero otherwise; ε is a normal random variable describing the variation in the trait not controlled by the analyzed QTL.[16] When analyzing a backcross, Equation 11.1 can be rewritten without the dominance component, as described in Equation 11.2:[16]

$$f_i = m + a \times B_i + \varepsilon \qquad (11.2)$$

Instead of using a linear model, a maximum likelihood (ML) method can be used to estimate the unknown parameters. While linear models are quite simple, they only take into account the marker mean and variance, while maximum likelihood methods use the full information from the distribution and therefore are expected to be more powerful. The disadvantage with ML methods is that they are computationally intensive and special programs are needed.

A.   Sample Size

The sample size required to detect linkage is influenced by the difference between marker class means and within marker class variance. If the sample size is increased, the within-class variance is reduced, which increases the power to detect linkage, while a change in experimental design is necessary to increase the difference between means. Another way to increase the power to detect linkage is to use a more elaborate phenotypic analysis including more explanatory variables and thereby reducing the residual variance. Lander and Botstein[17] have examined the power of maximum likelihood methods to detect linkage with interval mapping. They conclude that with at least one marker every 20 cM it is necessary to have a minimum of 250 *F2* backcross or *F2* intercross individuals to detect a locus that accounts for at least 5% of the variation.

B.   Selective Genotyping

Selective genotyping has been suggested[17,18] as a method to optimize resource usage in QTL mapping experiments. The basis for the strategy is that most of the linkage information resides in the individuals with an extreme phenotype. If it is much cheaper to score the trait than to score marker genotypes, the gain in power can be significant. It has been suggested that only the upper and lower 25% of the population will be scored. To correct for the sampling bias that is introduced when typing only the individuals with the extreme phenotypes, the unscored individuals must be included but with missing values for the genotypes,[17] or a special likelihood function that takes the bias into account can be constructed.[18]

C.   Statistical Significance

The balance between type I (the probability of a false positive result: a) and type II (the probability of a false negative result: b) errors is a common problem in the analysis of complex disorders. The power of a test is 1 – b, which is the ability to detect a linked QTL. When analyzing markers covering the entire genome, the large number of tests performed will increase the chance of a false positive result. To obtain an overall significance level, g, for n independent tests requires that each individual test has a significance level of:

$$\alpha = 1 - (1 - \gamma)^{1/n} \approx \gamma/n \qquad \textbf{(11.3)}$$

This is the *Bonferroni* correction for multiple comparisons. The assumption of independent tests is not valid in a whole genome scan with dense markers. Linked markers are clearly not independent, and if the standard Bonferroni correction is performed there is a loss of power to detect true associations.

If the power of an experiment is low and there is a large number of QTLs, the effect of the detected QTLs can be significantly overestimated. Lack of repeatability of an experiment may indicate low power. Permutation tests have been proposed[19] as a method to determine appropriate overall significance levels. Permutation tests also have the advantage of being model free and are therefore extremely robust. The test is performed by comparing the test statistics obtained in the analysis of the observed data with those generated under the hypothesis of no linkage. The latter are obtained by shuffling the phenotypic data while using the original genotypic data and then calculating the test statistics. This is done a large number of times to obtain an empirical distribution of values under the hypothesis of no marker trait association. A disadvantage with permutation tests is that they are numerically intense.

IV.   Gene Finding

After the identification of loci associated with disease, the next goal is to identify the responsible genes. This is certainly not an easy task. There are, in principle, three steps to reaching this goal:

- Isolation of loci by backcross breeding and a search for subphenotypes associated with the disease
- Identification and sequence analysis of candidate genes
- Transgenic constructions and biological testing of the importance of the gene

A.   Isolation of Loci by Backcross Breeding and Search for Subphenotypes Associated with the Disease

The gene mapping of *F2* intercross or backcross mice will reveal a big chromosomal fragment of 10 to 30 cM to which there is significant linkage. It is possible to decrease this fragment by extension of the number of animals in the *F2* analysis, but it is not realistic to reduce the region drastically by this method due to the high number of animals needed. In addition, the results will still be statistical and not absolute. A more secure but time-consuming way is to perform backcross breeding into one of the parentals and search for recombinations. Depending on the phenotype and breeding performance of the parentals, a choice can be made to isolate the resistant or susceptible allele. We have, for example, selected the C57Bl/10 (=B10) gene background as a parental strain, and in this strain we will isolate both resistant and susceptible loci. In the first backcross generation (BC2), we screen for the desirable recombinations. Different strategies for screening can be used.

First, a decision based on the balance between time and animal costs must be made. To find recombinations at a close distance, a large number of mice in the BC2 cross have to be screened (directly related to the cM distance). Alternatively, the screening can be made at each further generation of backcross. Depending on interacting genes at other loci, it might be important to have a control of the gene background. This will require screening for other loci or just further backcrossing until the genomic background is statistically of parental origin. It is also wise to have a strategy by which recombinant fragments can be isolated in the congenic strains. The required numbers of animals will be possible to estimate.

Second, a decision must be made concerning the phenotypes to be screened. The disease phenotype requires long-term experiments and is complicated to perform. Moreover, the selected gene might have low penetrance and will, therefore, require a very large number of animals to reach statistical power. It is very helpful to select a more simple subphenotype which might be associated with disease and which is linked to the same gene region. This will require testing of new phenotypes based on what is known about the particular disease and what is known about possible candidate genes in the selected DNA fragment. The establishment of congenic strains will provide a powerful tool for testing new phenotypes related to disease.

The goal for the backcross breeding is to establish a congenic strain carrying a gene responsible for the disease or a related well-defined phenotype, within a DNA fragment of less than 1 megabase. A region of this size is feasible to sequence to identify candidate genes.

B. Cloning of Candidate Genes

Isolation of candidate genes can be done using several methods. Linkage mapping and analysis of genetic recombinations in the region of interest will delimit the region to a small enough fragment suitable to sequence. There are a number of methods for isolating polymorphic genes in the delimited fragment, but they are not the subject of the present review. The methodologies, however, will change dramatically in the next few years as we approach a situation when almost all genes are identified as expressed sequence tags (ESTs), for which the partial sequence is known. Such ESTs, when positioned on the genetic and physical maps, will provide handles to identify genes of interest. The isolated genes are then analyzed for mutations associated with the disease.

C. Transgenic Approaches

To obtain conclusive evidence, it will eventually be essential to obtain a physiological and functional expression of transgenes which contain naturally occurring polymorphisms. To reach this goal, endogenous genes could be mutated by homologous recombination in embryonic stem cells (ES cells), with their subsequent introduction into recipient blastocysts, production of chimeric founder animals, and generation of mice expressing the transgene. Constructs should be used which contain single mutations that mimic the allelic difference. It is also necessary to inactivate the expression of the gene in question or to control the expression of it. It is in many cases an advantage to test various possibilities at this stage. To reach this goal, the Cre/lox system can be used. The product of the bacteriophage P1 Cre gene is a recombinase which catalyzes reciprocal recombination at specific 34-bp direct-repeat sequences, the lox (locus of recombination) sites. Genomic DNA between two lox sites is removed if Cre is expressed in a cell. If the lox sites have been introduced into the genome as part of a targeted knockout construct also harboring an engineered point mutation, it becomes possible to later remove the inactivating NeoR cassette flanked by the lox sites by Cre and thus introduce a point mutation into the endogenous gene. This can be achieved in the ES cell by transient transfection of Cre. The insertion of a NeoR-tk cassette into specific regions of the genome allow for the introduction of further mutations in a second round of electroporation. A construct harboring the desired mutation but lacking any selectable marker gene can then be transfected into ES cells carrying the NeoR-tk cassette. Cells which have undergone homologous recombination with the transfected DNA will lose the tk gene and thus survive negative selection in gancyclovir-containing medium. This strategy allows for the introduction of mutations that differ among the analyzed alleles.

The establishment of such "knock-in" mice will provide the possibility to prove definitely the importance of the gene for disease development. Such mice will also be very useful for studies of the role of the gene in the pathogenesis of the disease. Not the least, it will be invaluable for the development of new therapeutic strategies.

V. Gene Finding: The MHC Class II *ABp* Gene as an Example

Some years ago we started a genetic analysis of the CIA model by analyzing the role of MHC in more detail. We and others demonstrated a differential susceptibility to CIA among inbred and congenic mouse strains.[15,20,21] The congenic strains differed only in MHC and connected regions on Chr 17. Arthritis susceptibility correlated with the H-2^q and the H-2^r haplotypes. Since the CII autoimmune responsiveness also correlated with H-2^q and H-2^r, we assumed that the genes responsible controlled immune responsiveness which, at that time, was thought to be dependent only on MHC class II genes. We therefore sequenced the MHC class II genes from H-2q, p, and r and some related class II genes from wild mice.[22,23] We found that the difference between the class II genes in the susceptible H-2^q haplotype and the resistant H-2^p haplotype was only four amino acids, all of which were located in the peptide-binding region in the Aβ chain (corresponding to the DQβ in humans). However, the congenic fragment differing between q and p contains hundreds of other polymorphic genes which may affect arthritis susceptibility. To demonstrate the genetic association conclusively, we produced transgenic mice using a genomic construct of the entire *ABp* gene in which the few nucleotides that differed from those in the q allele were replaced.[24] This new construct was shown to be physiologically expressed in H-2p mice by fluorescence-activated cell sorting (FACS), immunohistochemistry, and by antigen presentation to q-restricted CII-reactive T-cell clones. Interestingly, the same gene, when controlled by strong viral promoters (CMV), also showed physiological expression, which demonstrates that the expression must be controlled by the endogenous Aα chain.[24] These transgenic mice develop arthritis and CII immune response to the same extent as H-2q mice, showing that the A class II gene is responsible for the major genetic association of CIA. To our knowledge, this is the only conclusive evidence that naturally polymorphic MHC class II genes are critical for development of autoimmune arthritis.

References

1. Holmdahl, R. et al., *Immunol. Rev.,* 118, 193, 1990.

2. Spinella, D. G, Jeffers, J. R., Reife, R. A., and Stuart, J. M., *Immunogenetics,* 34, 23, 1991.

3. Andersson, M., Goldschmidt, T. J., Michaëlsson, E., Larsson, A., and Holmdahl, R., *Immunology,* 73, 191, 1991.

4. Jansson, L. and Holmdahl, R., *Clin. Exp. Immunol.,* 94, 459, 1993.

5. Jansson, L. and Holmdahl, R., *Eur. J. Immunol.,* 24, 1213, 1994.

6. Hansson, I., Holmdahl, R., and Mattsson, R., *J. Neuroimmunol.,* 27, 79, 1990.

7. Hansson, I., Holmdahl, R., and Mattsson, R, *Clin. Exp. Immunol.,* 92, 432, 1993.

8. Hansson, I., Holmdahl, R., and Mattsson, R., *J. Neuroimmunol.,* 39, 23 1992.

9. Holmdahl, R., Jansson, L., Andersson, M., and Jonsson, R., *Clin. Exp. Immunol.,* 88, 467, 1992.

10. Holmdahl, R., Jansson, L., and Andersson, M., *Arthr. Rheum.,* 29, 1501, 1986.

11. Jansson, L. and Holmdahl, R., *J. Reprod. Immunol.,* 15, 141, 1989.

12. Mattsson, R., Mattsson, A., Holmdahl, R., Whyte, A., and Rook, G. A. W., *Clin. Exp. Immunol.,* 85, 41, 1991.

13. Jansson, L., Mattsson, A., Mattsson, R., and Holmdahl, R., *J. Autoimmunity,* 3, 257, 1990.

14. Holmdahl, R. et al., *Clin. Exp. Immunol.,* 62, 639, 1985.

15. Holmdahl, R., Klareskog, L., Andersson, M., and Hansen, C., *Immunogenetics,* 24, 84, 1996.

16. Lander, E. S. et al., *Genomics,* 1, 174, 1987.

17. Lander, E. S. and Botstein, D., *Genetics,* 121, 185, 1989.

18. Darvasi, A. and Soller, M., *Genetics,* 138, 1365, 1994.

19. Churchill, G. A. and Doerge, R. W., *Genetics,* 138, 963, 1994.

20. Wooley, P. H., Luthra, H. S., Stuart, J. M., and David, C. S., *J. Exp. Med.,* 154, 688, 1981.

21. Holmdahl, R., Jansson, L., Andersson, M., and Larsson, E., *Immunology,* 65, 305, 1988.

22. Holmdahl, R., Karlsson, M., Andersson, M. E., Rask, L., and Andersson, L., *Proc. Natl. Acad. Sci. USA,* 86, 9475, 1989.

23. Gustafsson, K., Karlsson, M., Andersson, L., and Holmdahl, R., *Eur. J. Immunol.,* 20, 2127, 1990.

24. Brunsberg, U. et al., *Eur. J. Immunol.,* 24, 1698, 1994.

25. Smith, B. D., Martin, G. R., Dorfman, A., and Swarm, R., *Arch. Biochem. Biophys.,* 166, 181, 1975.

26. Miller, E. J. and Rhodes, R. K., *Methods Enzymol.,* 82, 33, 1982.

27. Bernard, M. et al., *J. Biol. Chem.,* 263, 17159, 1988.

28. Cremer, M. A. et al., *J. Immunol.,* 146, 4130, 1991.

29. Trelstad, R. L., Kang, A. H., Toole, B. P., and Gross, J., *J. Biol. Chem.,* 247, 6469, 1972.

30. Reese, C. A. and Mayne, R., *Biochemistry,* 20, 5443, 1981.

Appendix A. Scoring Collagen-Induced Arthritis

1. The animals should be scored in a well-lit area. Scoring should be done close to eye level. If using both males and females, start with the males. Lift the mouse out of the cage and place it on top of a wire-cage lid. Test the strength of the paws by gently pulling the tail. Feeble paws indicate disease and should be examined closer. Lift the mouse by the tail until the underside of the hind paws is visible. It is easy to examine the underside of the paws while the forepaws are still on the cage lid. Alternatively, the mouse can be inspected carefully by holding it in one hand by holding the neck between the thumb and the forefinger and the tail under the little finger.

2. Inspect each paw carefully and note signs of arthritis on the scoring protocol. Use the description in Table 11.4 to assign the correct score to the joints. Indicate the areas which are inflamed on the figure depicting the paws (Figure 11.1). Also see Figure 11.1 for examples of active inflammation. Deformity remaining after the swelling has subsided is usually easy to discern from active inflammation.

3. Note inflammation in other parts, i.e., eyes, ears, or tail, or other pertinent information in the comment box.

4. When the animals have had severe arthritis for more than 2 weeks, deformity often appears. Indicate this with a D in the scoring protocol.

5. After all inflammation has subsided, deformity remains. This ranges from inability to spread the toes to stiff joints and new bone formation.

TABLE 11.4
Summary of the Simple and the Extended Scoring Protocols

Localization	Extended Score	Simple Score	I.D. Letter[a]	Relation (Extended–Simple)
Single				
One toe	1	1	T	1 = 1
One knuckle	1	1	K	
Combinations				
2, 3, 4, or 5 toes swollen	2, 3, 4, 5	2	T	2–10 = 2
2, 3, 4, or 5 knuckles swollen	2, 3, 4, 5	2	K	
Midpaw swollen	5	2	M	
Ankle swollen	5	2	A	
Midpaw and 1 toe	5 + 1 = 6	2	MT	
Midpaw and 2, 3, 4, or 5 toes	5 + (2,..,5) = 7,..,10	2	MT	
Ankle and midpaw	5 + 5 = 10	2	AM	
Involvement of All Three Parts of Paw				
At least swollen in ankle + midpaw + toe	5 + 5 + 1 = 11	3		11–15 = 3
At least swollen in ankle + midpaw + (2 to 5) toes	5 + 5 + (2,..,5) = 12,..,15	3		
Recovery				
Paw swollen and deformed	(1,..,15) D	(1, 2, 3) D		
Paw not swollen but deformed and stiff	0 D	0 D		
Total score per mouse	60	12		

[a] T= toe; M = midpaw; A = ankle/wrist; K = knuckle.

Appendix B. Collagen Preparation Procedures

Collagen Preparation

Arthritogenic CII is commercially available (e.g., from Chondrex, Elastin, or Sigma); however, if large quantities are needed or if there are specific demands on species source or purities, it is recommended that you prepare and purify it. The source material can be cartilage tissues or the Swarm rat chondrosarcoma (RCS). Cartilage tissues can be taken from xiphoid process of sternum or joints. Xiphoideus from rats and mice and articular cartilage from calves are the simplest source of cartilage collagen. For xiphoideus, cut and collect the outmost half; at least 200 mice are needed for one purification. In articular joints, pieces of cartilage are simply cut from

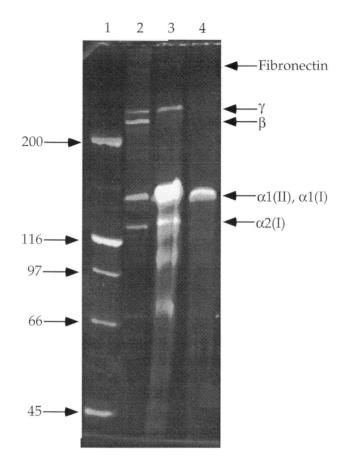

FIGURE 11.3

SDS-PAGE (4 to 12% gradient) of purified CII obtained from RCS after pepsin digestion (lane 4) and lathyrism (lane 3). Lane 1, high-range molecular weight standard in kDa (BioRad); Lane 2, CI obtained from pepsin digested rat tail (Sigma). All samples were reduced with 2-mercaptoethanol and collagens loaded in 40 µg.

knee or hip. Two different methods increase extractibility of collagen: pepsin digestion and lathyrism. Pepsin digestion cleaves short sequences in the carboxy and amino terminus portion containing intramolecular cross-links, thus solubilizing collagen molecules. Collagen from lathyritic tissues is obtained by feeding Swarm tumor transplanted rats or young animals with β-aminoproprionitrile monofumaratic salt (b-APN) *ad libitum* in water with a concentration of 3 g/l during growth of tumor. Formation of inter-cross-links between collagen molecules is blocked during growth, thus soluble collagen can be extracted.

In Figure 11.3, pepsin-digested and lathyritic CII was run on an SDS-PAGE. The bands were visualized with SYPRO-orange (BioRad), which has the same sensitivity

TABLE 11.5
Production of RCS

1. Suspend a part of the tumor in phosphate-buffered solution (PBS) by passing through a sieve.

2. Sedate the rat and do incisions caudally and laterally.

3. Use a long blunt-end metal needle for injection of 1 ml cell suspension in each side. Inject subcutaneously from the forward shoulder and along the entire side to the incision site. Close with an agraff.

4. Let the sarcoma grow for approx. 3 weeks. Inspect regularly and collect when a homogenous bump at the injection area is visible. If necrosis occurs in the incision site, the tumor must be collected immediately.

5. Kill the rat and collect the sarcoma by gently cutting and separating the tumor from skin. Keep the connective tissue surrounding the sarcoma until preparation; it will facilitate further handling.

6. Remaining tumor tissue is stored at $-20°C$, and a part is used to inoculate new rats.

as silver-staining. Both treatments show one distinct band of $\alpha 1(II)$. Compared with the CI standard in lane 2, both treatments have a band of $\alpha 2(I)$, which is in the ratio 10% of $\alpha 1(II)$ and $\alpha 1(I)$; β and γ indicate complexes of different α-chains, dimers and trimers, respectively. All preparations of collagen give some contamination of fibronectin.

Transplantation of RCS

RCS was first discovered as a spontaneous bone tumor arising in a Sprague-Dawley rat.[25] The tumor produces considerable quantities of cartilage-specific proteins, 40% of which consist of CII. The tumor must be maintained by transplantation in rats; strain and sex are of no importance (see Table 11.5)

Collagen Purification

All procedures should be performed at a temperature of 4°C. This minimizes bacterial growth, enhances solubility, and ensures the native conformation of collagen. All materials should be of glass, since collagen adheres fast to plastic. The essential steps in collagen purification have been described previously.[26-30] A purification protocol for pepsin-digested collagen is given in Table 11.6 and for lathyritic collagen in Table 11.7.

TABLE 11.6
Pepsin Digestion of Cartilage

1. Start with 100 to 200 g RCS or cut pieces of articular cartilage. Dissect the RCS from surrounding connective tissue that contains type I collagen (CI). For xiphoideus, start with 4 to 5 g and homogenize at low speed in dH_2O to remove connective tissue. Discard the water.

2. Cartilage or RCS is homogenized in 3 volumes by weight with guanidine buffer: 4 M guanidine-HCl and 50 mM Tris, pH 7.5. Extract proteoglycans by stirring the homogenate for 12 hr. Centrifuge at 23,000 g for 30 min; extract the pellet 2 more times. Keep the pellet or the xiphoideus. Further purification of xiphoideus, freezing with $N_2(I)$, and then grinding in a mortar are needed.

3. Resuspend the pellet or ground xiphoideus by weight with guanidine buffer and stir for 48 hr. Centrifuge as above and resuspend pellet in 200 ml dH_2O. Centrifuge and wash the pellet 3 more times.

4. Dissolve pepsin (originating from porcine stomach mucosa) to a concentration of 2 mg/ml in 0.5 M acetic acid. Add 0.5 g pepsin to 100 g tissue and incubate with stirring for 24 hr. Centrifuge as before and keep the supernatant. Redo digestion of the pellet. Pool the supernatants.

5. Inactivate pepsin by raising pH to 8.1 with NaOH. Centrifuge as before and keep the supernatant. Dialyze supernatant in tubing MWCO 12-14000, against 0.5 M acetic acid and 0.9 M NaCl in 5 l with stirring for 96 hr with at least 3 exchanges. Precipitate of collagen will be formed.

6. Centrifuge at 34,000 g for 40 min and dissolve pellet with 0.5 M acetic acid and approximately 1.5 ml/g original tissue. Dialyze the solvent in tubing, 5 × 2 hr, against 200 mM NaCl, 50 mM Tris-HCl (pH 7.4) in 5 l. Centrifuge as above and keep the supernatant.

7a. Type I collagen (CI) exclusion (optional): In xiphoid cartilage, there is a considerable amount of CI. CI contamination can be reduced, but with a great loss of CII. Measure the volume of the supernatant and add by dripping equal volume of 5 M NaCl solution slowly to supernatant with constant stirring. Stir for 24 hr. Centrifuge at 34,000 g for 40 min. Keep the supernatant.

7b. Pepsin/proteoglycan exclusion (optional): The contamination of dissolved pepsin and other negatively charged proteins can be reduced with the help of a cation exchange column. Equilibrate the column with 200 mM NaCl and 50 mM Tris-HCl, pH 7.4. Add sample and collect flow-through in fractions on one quarter supernatant volume. Continue to run equilibration buffer and collect fractions of one quarter supernatant volume. Pool viscous fractions containing collagen.

8. Add crystalline NaCl to the pooled fractions with stirring to a total concentration of 4.4 M NaCl. Let precipitate form without stirring for 24 hr.

9. Centrifuge at 34,000 g for 60 min. Dissolve the pellet in 0.5 M HAc and dialyze the solution in tubing, 4 × 2 hr, against 0.1 M acetic acid in 5 l.

10. Lyophilize the dialysate in a freeze-dryer. To obtain a fluffy-white, cottonlike product, rotation-freeze the dialysate at –70°C before freeze-drying. Measure the weight and dissolve to a concentration of 2 to 5 mg/ml in 0.1 M acetic acid. Store at 4°C; time will not affect the collagen under these conditions.

TABLE 11.7
Neutral Extraction of Lathyritic Tissue

1. Start with approx. 300 g and dissect lathyritic tissue from connective tissue.

2. Homogenize in a small volume of extraction buffer: 1.0 M NaCl and 50 mM Tris, pH 7.5. Add buffer to a total of 5 volumes of tissue weight. Extract collagen from the homogenate with stirring for 16 to 24 hr. Centrifuge at 23,000 g for 30 min; keep the supernatant. Extract the pellet once in 3 volumes of buffer, centrifuge as above, and keep the supernatant. Repeat with 2 volumes buffer. Pool the supernatants.

3. Add crystalline NaCl to the pooled supernatants to a total concentration of 4 M with stirring. Let precipitate form for 24 hr without stirring.

4. Resuspend the precipitate and centrifuge at 34,000 g for 30 min and keep the pellet. Dissolve the pellet in 0.5 M acetic acid until it is slightly viscous. Stir for 12 hr, centrifuge as above, and keep the supernatant.

5. Add crystalline NaCl to the supernatant to a concentration of 2 M with stirring. Let precipitate form for 24 hr without stirring. Resuspend and centrifuge as before and keep the pellet.

6. Dissolve the pellet in 0.5 M acetic acid. Dialyze the solution in tubing 5 times against 5 l of 200 mM NaCl and 50 mM Tris-HCl, pH 7.4. Centrifuge as before and keep the supernatant.

7. Cation exchange chromatography and further treatment as for pepsin-digested tissues.

Appendix C. Microscopic Scoring System for Collagen-Induced Arthritis

See Figure 11.2.

Evaluation of Arthritis Activity and Severity

Hind-paw sections or ankle sections are stained with hematoxylin-erytrocin (htx-e). All sections should be obtained at the same level through the specimens. If sections from the ankle joint are examined, they should be taken in the middle of the ankle joint.

Score 1: Mild synovitis with hyperplastic synovial membrane, small focal infiltration of inflammatory cells, increased numbers of vessels in synovium, and a villous formation synovium; no bone or cartilage erosions

Score 2: Moderate synovitis with pannus formation, bone and cartilage erosions limited to discrete foci, undisrupted joint architecture

Score 3: Severe pannus formation, with extensive erosions of bone and cartilage and disrupted joint architecture

If hind-paw sections are assessed, the number of affected joints with different scores can be added together, and the percentage of joints (based on total number of assessed joints) with different scores can be presented in a column diagram. The scoring should be performed in a "blinded" fashion. It is recommended to always start by examining specimen from normal control animals of the same strain (especially if score 1 is to be used).

Evaluation of the Healing Process of Arthritis

Hind-paw sections or ankle sections are stained with Safranin-O staining, visualizing cartilage (presence of proteoglycans). All sections should be obtained at the same level through the specimens. If sections from the ankle joint are examined, they should be taken in the middle of the ankle joint.

    Score I: New formation of cartilage; no visible new bone formation
    Score II: Cartilage and bone formation
    Score III: Cartilage and bone formation and apparent ankylosis

If hind-paw sections are assessed, the number of affected joints with different scores can be added together, and the percentage of joints (based on total number of assessed joints) with different scores can be presented in a column diagram.

Section V

Genome Mapping

Chapter

Human Genome Mapping

A. Collins and N. E. Morton

Contents

I.   Construction of High-Resolution Linkage Maps

Solomon and Bodmer[1] and Botstein et al.[2] first proposed the use of DNA polymorphisms as markers for linkage mapping, initiating a rapid and dramatic increase in the volume of linkage data available. The present state of the linkage map is reflected in recent publications, in particular, the GENETHON map of 5264 microsatellite markers[3] and the CEPH, version 8,[4] with linkage data for 10,975 markers, of which 7950 are microsatellites. Strategies for map construction have evolved to accommodate higher resolution mapping using increasingly powerful computers. In order to locate and subsequently clone and sequence disease genes, the main thrust of linkage mapping in recent years has been to construct maps of polymorphic DNA markers, the numbers of which far exceed those of expressed genes. Ultimately, however, the linkage map will become saturated with genes through the combination of physical and genetic mapping efforts. A number of computer programs for constructing maps of co-dominant markers have been in use for several years and have been applied extensively to, for example, CEPH family data.

The original DNA polymorphisms were restriction fragment-length polymorphisms (RFLPs), which are moderately polymorphic and require Southern blots and DNA probes. Throughout the 1980s (when these markers were by far the most abundant type), map construction, while relatively time-consuming even on the more powerful workstations, could be performed adequately with computer programs such as CRI-MAP, MAPMAKER, and MAP.[5] For n loci there are $n!/2$ possible orders, so evaluating the likelihood for more than a fraction of these in a map of 10 or more loci was not feasible. Strategies for constructing maps involved seriation, beginning with a reasonable starting order for a few loci, followed by various "build" approaches to insert loci into a "framework" map. Most of the multipoint programs such as CRI-MAP were used in such a way as to achieve support for an order at odds of 1000:1. Frequently this meant that many loci not supported at this level were dropped and the support status of others changed as other loci were added. With a "built" map, reliability of an order could be tested using local re-ordering strategies (for example, CRI-MAP FLIPS and the MAP bootstrap) which either permuted subsets of the loci or sequentially switched adjacent loci.

These methods were effective for the RFLP era, but the arrival of polymerase chain reaction (PCR)-based technologies and microsatellite markers has created a need for a different approach. For the first time, the generation of polymorphic markers and genotyping have become relatively simple, and as a result the rate of data acquisition has begun to outstrip the capacity of the analytical software. Recent developments in linkage mapping have therefore been concerned with algorithmic improvements, automation, and faster computer systems. There are two broad approaches to map construction: Multipoint mapping considers the likelihood for a number of loci simultaneously, while multiple pairwise methods are concerned with recombination between pairs of loci.

A.   Multipoint Mapping

There are a number of programs available for multipoint mapping of large numbers of markers. The program CRI-MAP[5] has been used extensively to construct maps of (relatively) small numbers of loci, but computational constraints have prompted the development of an automated method of map construction in the program *multimap* which uses CRI-MAP for computing likelihoods, accelerated in the parallel machine version, CRI-MAP-PVM.[6] The parallelized version has adapted the functions ALL and FLIPS of CRI-MAP. The ALL function takes a set of ordered markers and a set of unordered markers and evaluates the likelihood of all possible orders of the unordered sets in each interval and returns the most likely order. FLIPS is used to verify an order by permuting subsets of n loci sequentially along a map. Since all of the orders are permuted and stored prior to the calculation of likelihoods, this strategy is a good candidate for parallelization.

For the multipoint mapping of disease loci, the problem of handling increasing numbers of markers and marker alleles has led to serious computational problems. The standard approach implemented in the LINKAGE program[7] uses the Elston and Stewart[8] algorithm, which performs summation over all possible genotype vectors. Where an individual is untyped at several highly polymorphic markers, the number of possible genotypes becomes enormous. Strategies for accommodating this serious computational problem have included discarding partially informative data and splitting larger pedigrees into smaller ones. The VITESSE algorithm[9] addresses the computational problem for exact multipoint likelihoods in a number of ways, including more efficient memory usage and by genotype re-coding. This strategy reduces the number of genotype vectors by re-coding genotypes into sets of transmitted and nontransmitted alleles. The resulting improvement in efficiency is accompanied by improved mapping. The strength of the evidence for linkage at a specified recombination fraction, θ, is measured by an lod score. An lod is the log of the likelihood ratio $L(\theta)/L(0.5)$. O'Connell and Weeks[9] show that the lod score for the dopamine-responsive dystonia gene (DRD) to markers on chromosome 14 increases from 6.3 to 7.2 using VITESSE with six markers (only four markers after allele re-coding could be accommodated by LINKAGE).

B.   Multiple Pairwise Mapping

Multiple pairwise analysis is less powerful than multipoint mapping for construction of linkage maps in the absence of typing errors, but nevertheless has several advantages. Pairwise lods allow construction of maps that include both disease loci and markers without recourse to raw data. Multiple pairwise analysis allows for chiasma interference and is less sensitive to genotyping error than multipoint,[10] in which error simulates a double crossover, and under these conditions it has been shown to recover the correct marker order more frequently than multipoint analysis.

A model for error filtration[10] further reduces the impact of typing errors on the map. The MAP program[11] has been used extensively. The MAP algorithm requires the conversion of the recombination fraction θ and the corresponding lod (Z) into binomial estimates of numbers of meioses (N) and recombinants (R). The log likelihood is $\ln L = R (\ln \theta) + (N - R) \ln (1 - \theta)$, where θ is the expected recombination fraction. Iteration, for a given locus order, is on map distance (w) in centimorgans (cM) which is related to θ through a mapping function. The Rao map function[12] allows for a variable level of interference (through a parameter, p, where $1 - p$ measures the strength of interference).

C.   The MAP+ Program

MAP+[13] was developed for multiple pairwise linkage mapping of several hundred markers per chromosome, as in the more recent CEPH releases. The program has enhancements and features not available in the earlier MAP program.[11] MAP+ is closely associated with the enhanced genetic location database (*ldb+*) which provides starting (trial) linkage maps. Generation of sex-specific pairwise lods from CEPH genotype files that have been formatted as CRI-MAP genotype (.gen) files is achieved using a local version of CRI-MAP entitled CEPH2MAP which implements a sex-specific version of the TWOPOINT function to give lods factored by sex. Files of lods (partial lod files) from CEPH, LODSOURCE, and the literature are stored in the location database and are merged into a single datafile (*lod*).

Trial maps used in MAP+ are generated by the *ldb+* software from the best currently available summary map of the chromosome. Loci without a summary map location are assigned the same trial location as the marker with the highest lod. The emphasis of MAP+ is on improvement of a dense but locally unreliable trial map by maximizing the multiple pairwise likelihood.

MAP+ implements a novel locus-oriented algorithm to ordering loci which is extremely rapid even in maps of more than 100 loci and makes simultaneous analysis of maps of several hundred loci feasible on powerful workstations. This method finds the location for a locus that maximizes the log likelihood while keeping the locations in the remainder of the map fixed. This permits movement of a marker outside of its flanking markers and at convergence gives a standard error on location conditional on the remainder of the map. This process is applied to all loci in the map and is continued until no further improvements in log likelihood are obtained. Locus ordering is followed by iteration of intervals for a fixed order. MAP+ recovers the correct locus order and distance even for a very poor starting map.[13] The reliability of the order may be examined using a simplified version of the MAP bootstrap which evaluates the likelihood when the adjacent loci are switched. The MAP+ program also estimates the level of interference through the Rao mapping parameter and the typing error frequency, ε,[10] and implements methods for nondisjunction and radiation hybrid mapping.

II.   Mapping Complex Traits

Positional cloning techniques have enabled the genetic mapping of more than 500 genes involved in human diseases with simple Mendelian inheritance. While mapping of such traits is now relatively straightforward, the challenge of mapping complex traits (which quite often are "common" diseases such as heart disease, asthma, and diabetes) is only just being met. Complexity arises when the genotype:phenotype correspondence breaks down. This happens when the same phenotype occurs with different genotypes or different phenotypes have the same genotype. This may be due to environmental effects, incomplete penetrance, genetic heterogeneity, or multiple interacting genes. With heterogeneity, mapping of the genes involved is made more difficult because some families will show co-segregation of a chromosomal region with the disease and others will not. In polygenic inheritance, there is greater difficulty because a number of genes of small individual effect may be involved, greatly reducing the possibility of detecting particular genes. There is thus a spectrum of complexity from monogenic inheritance through multilocus inheritance, where dependent major genes are involved (heterogeneity), to multilocus inheritance where there may be interaction. Individuals with a predisposing genotype often do not have the disease (low penetrance), while individuals who lack a predisposing genotype still have the disease through environmental or chance effects (phenocopies). Furthermore, while affection for some traits, such as type I diabetes, may be unequivocally defined, others (such as asthma) have a less sharply defined phenotype and a spectrum of severity which is more appropriate as a quantitative trait rather than a qualitative (disease or no disease) classification.

While there has been some success in mapping genetically heterogeneous traits with major genes (for example, breast cancer and hereditary non-polyposis colon cancer), few genes involved in polygenic traits have been identified. An example of a polygenic trait for which some progress has been made is insulin-dependent or type 1 diabetes, IDDM.[14,15] IDDM is the most serious form of diabetes, an incurable autoimmune disease affecting 1 in 300 to 400 children.[15] For this complex trait, genetic mapping is simplified by straightforward diagnosis, early onset, and the use of animal models. A recent whole-genome analysis has built upon work which, through a candidate gene approach, had already identified two regions of the genome that are implicated in IDDM (the major histocompatibility complex on 6p21 called IDDM1 and the insulin gene region of 11p15 or IDDM2). The analysis implicated up to 11 other regions.[14] The long and difficult processes of confirmation of many of these linkages, unraveling the interactions between genes, and positional cloning have just begun and are hampered by the problems outlined above.

Most of the serious monogenic disorders are associated with a high risk to other family members. This can be parameterized as λ, the risk to relatives of a proband vs. the general population prevalence. For common diseases such as schizophrenia or diabetes, typical values for the relative risk among siblings (λ_s) are in the range of 3 to 20 (15 for type 1 diabetes). For a monogenic disorder, λ may be much higher (e.g., 500 for cystic fibrosis). The smaller the value of λ_s, the more difficult it is to locate the gene(s) involved and the less useful positional cloning is likely to be. It is usually not feasible to map through linkage a disease gene with λ less than 3.

A.   Parametric Methods

While most recent developments have been in nonparametric approaches to analysis, parametric methods which have been used extensively for Mendelian disorders are potentially very useful for complex traits. Such methods are parametric in the sense that they are concerned with estimating disease gene frequencies, age-specific penetrances, and mode of inheritance of the trait. Simultaneous segregation and linkage analysis fits a general model to the inheritance pattern of a trait. The majority of segregation analyses use families selected through affected individuals (probands) and must include correction for ascertainment bias in the sampling scheme. The models implemented in the computer programs POINTER[16] and COMDS[17] use nuclear families (parents and their children) together with pointers (a pointer is an affected relative outside the nuclear family through whom the family was selected) to control ascertainment bias. The likelihood of the phenotypic data under a series of models is maximized. The POINTER program implements the mixed model (mixed in that it includes a major locus and polygenes) and includes a quantitative trait but not a linked marker. Concerns about violation of distributional assumptions with quantitative traits and the need to include a linked marker led to the development of the COMDS program for two disease loci, a major linked locus together with a second (modifier) locus. The program analyzes a dichotomous trait but includes information on complex phenotype where there are additional phenotypic indicators of (genetic) severity among affected or predisposition (diathesis) among normals. The parameters of the major locus are d, the degree of dominance; t, the distance on the liability scale between the two homozygote means; q, the disease gene frequency; the recombination fraction between the trait and the marker locus, th; and information on complex phenotype parameterized as S (severity) and B (diathesis). Corresponding parameters (dm, tm, qm, Sm, Bm) are defined for the unlinked modifier locus. Additivity on a probit or logit scale is assumed. Among alternative programs, the class D model of *regress*[18] corresponds to the mixed model for a quantitative trait, but to no genetic model for a qualitative trait. It does not allow for linkage or ascertainment. Programs such as LINKAGE can fit a simple pre-assigned model but cannot estimate its parameters.

The "correct" parametric model must be more powerful for the detection of linkage than a nonparametric technique.[19] Segregation analysis by itself can give good estimates of genetic parameters for a major locus but is worthless for polygenes. Combined segregation and linkage analysis is more promising, but is a two-locus model more powerful than a nonparametric method for the detection of linkage in complex inheritance? Simulation offers a possibility for evaluating characteristics of different approaches, but the range of potential models to test is enormous. Perhaps only by application to real data will the relative merits of these alternatives be demonstrated.

A recent application of COMDS to Graves disease[20] produced a considerable increase in the lod score for linkage to HLA-DR3 (4.65 to 5.67) when residual heritability was modeled as a modifier locus. Additional information on a complex phenotype (thyroid antibody levels amongst unaffected individuals) increased the lod further to 6.64. This illustrates the potential increase in power to detect linkage with a two-locus model, particularly when modeled as a complex phenotype. A quantitative trait may be covariance-adjusted for known genotype effects when other loci are sought.

B.   Nonparametric Methods

A parametric model may not be able to distinguish between the many possible modes of inheritance for a complex trait. The two-locus model with a linked marker described above works well for certain traits, but presumably would be less effective for traits with a number of interacting genes of relatively small individual effect. Furthermore, ascertainment biases may be difficult to fully account for in the model. Nonparametric models, in particular allele-sharing methods, offer an alternative strategy for mapping complex traits. These models most often consider pairs of relatives and can accommodate quantitative or qualitative traits. Frequently, for a qualitative trait, the analysis is restricted to affected pairs. The approach seeks to demonstrate that the inheritance pattern of a chromosome region is not consistent with Mendelian segregation among affected relatives and that this region is shared by affected relatives more often than would be expected by chance. Nonparametric models are robust to the presence of phenocopies, incomplete penetrance, and sampling biases.

The majority of allele-sharing methods, for which there is now a large and growing literature, deal with pairs of affected sibs. Restricting the analysis to affected pairs has the advantage of reducing the number of individuals that must be typed with polymorphic markers, an important consideration in a whole-genome screen. Risch[21] developed this method by considering identity by descent (IBD) for a pair (the probability that they receive copies of the same *ancestral* alleles). This should be distinguished from identity by state (IBS), which is concerned with two individuals having the same alleles at a marker. This method has been used with sparser maps and less polymorphic markers where IBD status could not be inferred. IBS methods, based on means and regressions, require arbitrary weights, to which the results are highly susceptible. A better approach is through lods that are functions of IBD and gene frequencies only.

The power for detecting linkage by IBD depends on the risk ratio λ and the recombination fraction between marker and the trait. The risk ratio λ decreases with the degree of relationship between the proband and relatives, and the rate of decrease depends on the mode of inheritance underlying the trait.[21] For monogenic or additive multilocus inheritance, the $\lambda - 1$ ratios decrease by a factor of two for each degree of relationship. For a disease in which there is epistasis, the ratios decrease by more than a factor of two for each degree of relationship.

Many of the recent developments have been concerned with improving the performance of pair-based tests in terms of power (by constraining to "biological possibility", extending to other relatives or multipoint methods) and by reducing the type 1 error (see, for example, Lander and Kruglyak[22]). The region of biologically possible models for affected sibling pairs was first described by Holmans,[23] who determined that sharing probabilities (z_i, where $i = 0$, 1, or 2 alleles IBD) could be expressed in terms of a possible triangle into which all genetic models must fall. For a completely recessive, fully penetrant disease (where there is zero recombination with the trait), z_i values for an affected pair are 0, 0, 1, since both alleles must be shared IBD. With no linkage there is the expected Mendelian segregation (0.25, 0.5, 0.25) and for a fully dominant trait, 0, 0.5, 0.5. Holmans' development of the Risch method was to constrain sharing probabilities to the possible triangle with a consequent increase in power.

Morton[19] developed the β model which is appropriate to all pairs of relatives for either quantitative or qualitative traits. Recurrence risks are modeled by a single parameter (β). The relative recurrence risks are

$$\text{Parent-offspring, } \lambda_o = e^\beta$$

$$\text{Sibs, } \lambda_s = (1 + e^\beta)^2 / 4$$

$$\text{Monozygous twins, } \lambda_m = e^{2\beta}$$

For affected pairs, Risch[21] defined the sharing probability as:

$$Z_0 = 0.25 / \lambda_s$$

$$Z_1 = 0.5\, \lambda_o / \lambda_s$$

$$Z_2 = 0.25\, \lambda_m / \lambda_s$$

Under the β model, these can readily be expressed as:

$$Z_0 = 1 / (1 + e^\beta)^2$$

$$Z_1 = 2\, e^\beta / (1 + e^\beta)^2$$

$$Z_2 = e^{2\beta} / (1 + e^\beta)^2$$

For a pair of relatives scored Y_1, Y_2 for a normally distributed quantitative trait and x for IBD at a co-dominant marker, the correlation between the squared difference in trait score, $f = - (Y_1 - Y_2)^2$, and x provides a one-sided test of linkage.[24] This follows the method first proposed by Haseman and Elston.[52] Morton[19] has suggested, however, that f in the β model be defined as $f = Y_1 Y_2 / U$, where U is the mean for affected pairs. This gives more weight to pairs of relatives in the tails of the distribution, following Risch and Zhang,[26] who established that the majority of sibling pairs offer little power to detect linkage. Only pairs concordant for high trait values, low trait values, or extremely discordant pairs (from the lower and upper deciles) provide substantial power. This is an important consideration as, even with multipoint analysis, thousands of sibling pairs are required to detect linkage to a locus with a heritability of 25%.[27] The Risch and Zhang approach places the emphasis on highly selected sibling pairs rather than blanket genotyping. As an illustration for random simulated sibling pairs, analyzed with the Haseman-Elston[25] method for quantitative traits, 1082 pairs are required for significance, compared to only 61 when the most discordant pairs are used. While there may be practical difficulties in obtaining family material that satisfies the selection criteria, this sampling scheme offers a significant potential increase in power with a marked reduction in the volume of genotyping required when looking for linkage to a quantitative trait.

Like the β model, the γ model[28] also specifies one parameter but constrains $z_1 = 0.5$ for any value of γ. Collins et al.[29] have demonstrated that this has several disadvantages, all of which are addressed by the β model. In particular, the γ model does not model multiplicative effects on penetrance of disease loci, does not extend to relatives other than siblings, and fits available data on complex traits less well. The β model was shown to be more powerful than nonparametric alternatives, but there has been little comparison with parametric tests.

Krugylak and Lander[28] have extended sibling pair analysis to a multilocus test in the program MAPMAKER/SIBS. This has the advantage of using all the genetic markers to extract the IBD status at each point along the chromosome using an "inheritance vector" in which each meiotic event is coded 0 or 1 depending on whether the allele is paternally or maternally derived. As, usually, the inheritance vector cannot be uniquely described by the data, the program uses a hidden-Markov algorithm for computing the probability distribution of inheritance vectors. The increase in power to detect linkage is impressive. For IDDM, the lod at HLA increases from 8 to 11 when compared to the single point test. While MAPMAKER/ SIBS implements the γ model, this program currently is being extended to include the β model and will be available from the genetic location database World Wide Web (WWW) site.

The question of how to weight dependent sibling pairs has recently been addressed.[24] A sibship of s siblings contributes $s(s - 1)/2$ pairs if all possible pairs are considered. All possible pairs of siblings are required for an efficient test, even though there are only $s - 1$ independent sibling pairs. Thus, there is redundancy which increases with sibship size, so a number of weighting schemes have been considered. Wilson and Elston[30] proposed that each sibling pair be given equal weight but that the value of $s - 1$ be used to determine the degrees of freedom contributed by a sibship. Extensive simulation tests[24] have shown that pairs should be equally weighted and that the Fisher Z (r) transform (with infinite degrees of freedom) is a more reliable test statistic; however, it seems likely that nonparametric methods based on means and regressions will soon be superseded by lods using the β model.

III.   Map Integration

A.   The Genetic Location Database

An integrated map is a synthesis of a potentially large number of partial maps, each of which contains a subset of genetic and/or physical locations of loci. Morton and Collins[11] began the development of a location database (*ldb*) in which, for each chromosome, there is an integrated summary map that describes each locus by a vector of physical and genetic locations. This vector comprises cytogenetic, radiation hybrid, physical, genetic, and mouse homology data together with a composite location which gives an estimated physical location (measured in megabases from the p telomere) for the locus. The summary map is ordered by composite location. The simplest and most effective approach to constructing the composite location was found to be a priority rule: physical > radiation hybrid > genetic > cytogenetic data. This rule derives composite locations from physical data where that particular field is not blank for the locus, followed by data from radiation hybrids and so on. Scaling of the composite location is maintained by interpolation between closest flanking markers that already have composite locations, including telomeres and the centromere.

Cytogenetic locations in megabases are obtained as band midpoints from a scaled *band* file[31] that gives the location of cytogenetic band borders at 850-band resolution in megabases from the p telomere, assuming known chromosome arm length.[32]

The location database, in addition to summary maps for each chromosome and *band* files, contains fully referenced partial maps from the literature or personal communication (p-files) and pairwise lod files (L-files) containing lods from CEPH and the LODSOURCE database which stores lods for genes from the literature. The L-files provide the raw data for linkage mapping with the MAP+ program. The *arm* file provides the best current available estimates of chromosome arm lengths on physical and genetic scales, and the *alias* file contains a list of discontinued symbols and the current locus designation. There is also provision for raw radiation hybrid data (r-files) and clonal data (c-files).

The location database is a flat file system supported by the *ldb+* software[33] which performs many functions within the database: formatting files, updating symbols by reference to the *alias* file, sorting on specific fields, scaling where chromosome parameters have been updated, and constructing a list of references for partial maps and lod files. In addition, *ldb+* provides functions for updating the summary map with addition of new data. There are three main functions. *Incorporate* adds data by interpolation into a summary map where the locus is missing or there is a blank for that field. *Rectify* retains all of the orders and locations in the partial map and updates the summary map followed by interpolation of the original summary locations. This function is appropriate where a partial map is considered to provide a particularly reliable set of locations (both order and distance), which should therefore take priority. The *neighbor* function updates the summary map by obtaining a weighted mean location, using all partial maps, from the distance to upper or lower flanking markers. More weight is given to dense maps in this scheme which tends to favor more recent and more reliable partial maps. The relatively simple structure of the location database is particularly appropriate to the Internet and particularly the WWW, which allows locus-oriented, hypertext-based searching of the entire database (see Section IV, Mapping Resources).

B.   Map Projection

The location database maintains scaled maps both physical and genetic. Chromosome arm lengths are scaled to the physical estimates[32] and the genetic estimates.[33] Difficulties with integrating partial maps arise through discontinuities and distance errors in the physical scale. Partial physical maps cover a relatively small segment of the chromosome and are of uncertain location and size, and perhaps orientation. As data accumulate and the map becomes more complete, the orientation of a segment will be resolved by other data, either genetic or physical. Although there are now high-resolution (if not fully connected), YAC-STS-based physical maps covering a large fraction of the genome,[34] distances on these maps are inaccurate. Improvements in physical maps through measurement of YAC sizes, FISH, and other methods will be continued as sequencing approaches connectivity.

In the interim, an alternative strategy for establishing a physical scale has been developed.[33] A crude physical location is obtained, where known, from a cytogenetic band midpoint and the location estimated from the *band* file. Then a projected genetic map is obtained by fitting a double logistic curve to each chromosome arm, giving more weight to assignments to narrow bands. The genetic map projected onto a physical scale gives a sparse set of anchor markers which provide connectivity and scaling to order-only physical maps and maps of relatively small segments.

C.   A Metric Map of the Human Genome, Recombination, and Chromosome Structure

Recent work[33] has used these provisional whole genome integrated maps to examine properties of the genome at the level of Giemsa bands (850-band resolution). Each locus was assigned its most likely cytogenetic band from composite location and genetic location by interpolation between anchor markers. Regression models were fitted to a number of variables of biological interest (recombination density per megabase, female/male recombination rate, and locus density on the megabase and centimorgan scales). Analysis of these properties by band indicated the importance of positional effects. Telomeric bands are associated with high recombination, a low female/male recombination rate (since male recombination rate is relatively higher towards the telomeres), and high locus density. The heterochromatic regions showed correspondingly low values for these properties. High recombination density was very significantly associated with high marker/gene density as might be expected for a process for which the function is to recombine genes. Surprisingly, non-telomeric R (reverse-staining Giemsa) bands were not significantly different from G (Giemsa) bands. This may be because R bands at the 850-band level are poorly characterized with respect to GC content or that band composition is confounded with gene density and positional effects.

D.   Map Integration and the Human Genome Project

Although developed as a research tool and with limited resources, the location database remains the only system specifically designed to integrate maps in which the evidence that contributes to a location is retained and fully referenced. While it was hoped that the system or at least the fundamentals of the system might be adopted by the genome community, there has been a strong drift towards graphical displays in which the raw data are lost or obscurely encoded. This can be a useful laboratory tool for detecting gaps and errors in high-resolution mapping, and such a system (ACeDB) has been successfully applied to the nematode genome;[5] however, it is disastrous for map integration.

The SIGMA program[5] has been used to construct consensus maps at single chromosome workshops. While consensus maps may have been acceptable in the early days of the mapping effort when small numbers of loci were involved, this approach has rapidly become obsolete as the resolution of the human genome has increased by 45% per annum during the last 10 years. The single chromosome workshops have not been effective at producing integrated and reliable chromosome maps. Consensus maps lack supporting evidence and are not infrequently discarded when the workshop report is published and a new consensus map is constructed at the next meeting. There has been no agreement on standards for map presentation that would permit capture of quantitative data behind subjective maps.

At the urging of the Human Genome Organization (HUGO), during the last decade the Human Genome Project has relied on the Genome DataBase (GDB) for information support and has devoted much time and money to data entry and system development. Unfortunately, the relational database structure is unfavorable to map integration and more suited to small question/small answer queries. The problem faced by GDB has been clearly stated:[5] "Essentially, GDB represents a classical management dilemma. An ill-defined problem arises and an inherently limited solution is adopted; then as a consequence the existing investment becomes a compulsive reason for yet more investment." The WWW has made the situation worse for such a system. As the Web has become the essential tool for data and information acquisition through hypertext links, the complex software defining the GDB interface has become obsolete. Efforts to accommodate the WWW in GDB version 6 have not been well received. While the previous version (5.7) stored some partial maps for different chromosomes, version 6 has adopted a graphical display recently described as "too little, too late".[35] Virtually none of the information that belongs in a location database can be accessed through GDB, and the descriptive material is presented less well than mouse and *Drosophila* catalogs. A way must be found to create a database worthy of the costly information on which positional cloning and sequencing depend. Failing that, the major sources of information are listed below.

IV.   Mapping Resources (CD-ROM and Internet)

- *Genome Interactive Databases (GID):* A consortium of interactive databases on the human genome mapping. This CD-ROM is published by John Libbey Eurotext, 127 av de la République, 92120 Montrouge, France. The GID provides a hypertext interface to a number of databases (GENATLAS, CEPH, Généthon, *ldb+, * GENSET).
- *Généthon:* Available on the WWW at

    http://www.genethon.fr/genethon_en. html/.

- *CEPH:* Available on the WWW at

    http://www.cephb.fr/bio/.

    Includes CEPH genotype database and CEPH-Généthon physical mapping data.
- The Genome Database (GDB): Available on the WWW at

    http://gdbwww.gdb.org.

- *The Whitehead Institute for Genome Research:* Available on the WWW at

    http://www.genome.wi.mit.edu/.

    Contains data for the physical (YAC-STS) mapping project, radiation hybrid data, and software including GENEHUNTER for multipoint linkage analysis in pedigrees and MAPMAKER/SIBS for multipoint linkage with sibling pairs.
- *The genetic location database (ldb+):* Available on the WWW at

    http://cedar.genetics.soton.ac.uk./public_html/.

    Provides integrated maps of the human genome with locus-driven, hypertext links that generate a report from the location database providing links to the genome database (GDB) and to all partial maps that contain the locus. Also listed are the ten largest lod scores and locus content of clones that include the locus. Software available from this site includes COMDS, POINTER, MAP, MAP+, *ldb+,* and CEPH2MAP.
- *Genetic linkage analysis web server:* Available on the WWW at

    http://linkage.cpmc.columbia.edu/.

    Many programs are available for this site, including CRI-MAP, LINKAGE, CRI-MAP-PVM, *multimap,* SIGMA, VITESSE, and *Regress.*
- *UK MRC Human Genome Mapping Project Resource Centre:* Available on the WWW at

    http://www.hgmp.mrc.ac.uk/.

- *The Cooperative Human Linkage Center (CHLC):* Available on the WWW at

    http://www.chlc.org/.

References

1.   Solomon, E. and Bodmer, W. F., *Lancet,* i, 923, 1979.

2.   Botstein, D., White, R. L., Skolnick, M. H., and Davies, R. W., *Am. J. Hum. Genet.,* 32, 314, 1980.

3.   Dib, C., Fauré, S., Fizames, C., Samson, D., Drouot, N., Vignal, A., Millasseau, P., Marc, S., Hazan, J., Seboun, E., Lathrop, M., Gyapay, G., Morrissette, J., and Weissenbach, J., *Nature,* 380, 152, 1996.

4.   Dausset, J., Cann, H., Cohen, D., Lathrop, M., Lalouel, J.-M., and White, R., *Genomics,* 8, 575, 1990.

5.   Bishop, M. J., Ed., *Guide to Human Genome Computing,* Academic Press, London, 1994.

6.   Matise, T. C., Shroeder, M. D., Chiarulli, D.M., and Weeks, D. E., *Hum. Hered.,* 45, 103, 1995.

7.   Lathrop, C. M. and Lalouel, J. M., *Am. J. Hum. Genet.,* 36, 460, 1984.

8.   Elston, R. C. and Stewart, J., *Hum. Hered.,* 21, 523, 1971.

9.   O'Connell, J. R. and Weeks, D. E., *Nature Genet.,* 11, 402, 1995.

10. Shields, D. C., Collins, A., Buetow, K. H., and Morton, N.E., *Proc. Natl. Acad. Sci. USA,* 88, 6501, 1991.

11. Morton, N. E. and Collins, A., *Ann. Hum. Genet.,* 54, 235, 1990.

12. Rao, D. C., Morton, N. E., Lindsten, J., Hulten, M. C., and Yee, S., *Hum. Hered.,* 27, 99, 1977.

13. Collins, A., Teague, J., Keats, B. J., and Morton, N. E., *Genomics,* 36, 157, 1996.

14. Davies, J. L., Kawaguchi, Y., Bennett, S. T., Copeman, J. B., Cordell, H. J., Pritchard, L. E., Reed, P. W., Gough, S. C. L., Jenkins, S. C., Palmer, S. M., Balfour, K. M., Rowe, B. R., Farrall, M., Barnett, A. H., Bain, S. C., and Todd, J. A., *Nature,* 371, 130, 1994.

15. Todd, J. A., *Proc. Natl. Acad. Sci. USA,* 92, 8560, 1995.

16. Morton, N. E., Rao, D. C., and Lalouel, J.-M., *Methods in Genetic Epidemiology,* Karger, New York, 1983.

17. Morton, N. E., Shields, D. C., and Collins, A., *Ann. Hum. Genet.,* 55, 301, 1991.

18. Bonney, G. E., Lathrop, G. M., and Lalouel, J.-M., *Am. J. Hum. Genet.,* 43, 29, 1988.

19. Morton, N. E., *Proc. Natl. Acad. Sci. USA,* 93, 3471, 1996.

20. Shields, D. C., Ratanchaiyavong, S., McGregor, A. M., Collins, A., and Morton, N. E., *Am. J. Hum. Genet.,* 55,540, 1994.

21. Risch, N., *Am. J. Hum. Genet.,* 46, 242, 1990.

22. Lander, E. and Krugylak, L., *Nature Genet.,* 11, 241, 1995.

23. Holmans, P., *Am. J. Hum. Genet.,* 52, 362, 1993.

24. Collins, A. and Morton, N. E., *Hum. Hered.,* 45, 311, 1995.

25. Haseman, J. K. and Elston, R. C., *Behav. Genet.,* 2, 3, 1972.

26. Risch, N. J. and Zhang, H. P., *Am. J. Hum. Genet.,* 58, 836, 1996.

27. Fulker, D. W. and Cardon, L. R., *Am. J. Hum. Genet.,* 54, 1092, 1994.

28. Krugylak, L. and Lander, E. S., *Am. J. Hum. Genet.,* 57, 439, 1995.

29. Collins, A., MacLean, C., and Morton, N. E., *Proc. Natl. Acad. Sci. USA,* 93, 9177, 1996.

30. Wilson, A. F. and Elston, R. C., *Genet. Epidemiol.,* 10, 593, 1993.

31. Morton, N. E., Collins, A., Lawrence, S., and Shields, D. C., *Ann. Hum. Genet.,* 56, 223, 1992.

32. Morton, N. E., *Proc. Natl. Acad. Sci. USA,* 88, 7474, 1991.

33. Collins, A., Frezal, J., Teague, J. W., and Morton, N. E., *Proc. Natl. Acad. Sci. USA,* 93, 14771, 1996.

34. Hudson, T. J., Stein, L. D., Gerety, S. S., Ma, J., Castle, A. B., Silva, J., Slonim, D. K., Baptista, R., Kruglyak, L., Xu, S.-H. Hu, X., Colbert, M. E., Rosenberg, E., Reeve-Daley, M. P., Rozen, S., Hui, L., Wu, X., Vestergaard, C., Wilson, K. M., Bae, J. S. et al., *Science,* 8270, 1945, 1995.

35. van Heyningen, V., *Nature Genet.,* 13, 134, 1996.

Chapter

Somatic Cell Hybrid Mapping Panels: Resources for Mapping Disease Genes

Joan Overhauser

Contents

I.   Introduction

Current approaches for the physical mapping of the human genome utilize multiple methods for the dissection and characterization of human chromosomes. Somatic

0-8493-4411-5/98/$0.00+$.50
© 1998 by CRC Press LLC

cell hybrid isolation and manipulation represent one method that will be instrumental to achieving the goal of mapping and sequencing the human genome. In addition to their usefulness as mapping resources, the isolation of somatic cell hybrids can provide important information about the location of the critical chromosomal regions that are involved in aneuploidy syndromes.

Somatic cell hybrids are the result of fusing cells from two different species. For human gene mapping, a human cell is typically fused with a rodent cell line (mouse or hamster). After fusion, a heterokaryon is produced with representative chromosomes from both species. For some yet unknown reason, the human chromosomes are unstable and will not be replicated as efficiently as the rodent chromosomes. As a result, the human chromosomes are randomly lost. Eventually, a set of human chromosomes is stably replicated in the cell; therefore, the resulting clones of cells from a fusion experiment are a mixture of cells with different complements of human chromosomes. Through the use of auxotrophic rodent cell lines and the use of selective conditions, cell hybrids that have retained a particular human chromosome can be specifically selected.[1] Additional human chromosomes that are nonselectively retained are very common in the somatic cell hybrids.

In this chapter, methods for the construction of somatic cell hybrids, strategies for screening somatic cell hybrids, and the way in which sets of somatic cell hybrids are used to localize disease genes relative to patient samples are described. The techniques described here are directly applicable to any other human chromosome as long as a rodent cell line that can be used to select for the human chromosome is available. To demonstrate the usefulness of somatic cell hybrids in disease analysis, experiments using samples that are derived from patients with the 18q- syndrome are discussed.

II.   Construction and Screening of Somatic Cell Hybrids

For the efficient isolation of somatic cell hybrids that have retained a chromosome of interest, it is necessary to have available an auxotrophic cell line that allows for the selection of hybrids that have retained a specific human chromosome homologue. In the case of chromosome 18, two such rodent cell lines are available: a temperature-sensitive Chinese hamster cell asparaginyl-tRNA synthetase mutant (UCW206)[2] and a thymidylate synthetase-deficient Chinese hamster cell line.[3] The genes that will complement both mutant genes map to chromosome 18.[2,3] For this chapter, protocols using the asparaginyl-tRNA synthetase mutant cell line will be described. With this mutant cell line, only a change in temperature is needed to select for the human gene. With most other rodent mutant cell lines, specific media are used to grow the rodent cell line and the resulting selected hybrids. A flowchart for the isolation and screening of somatic cell hybrids is shown in Figure 13.1.

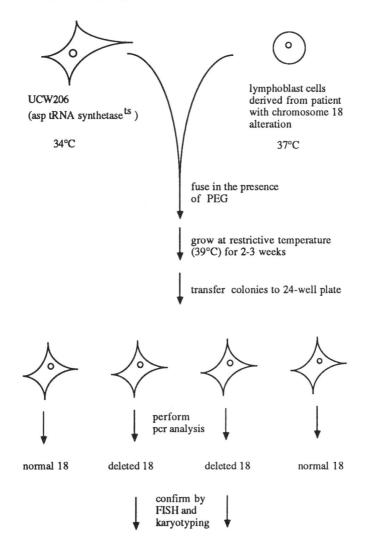

FIGURE 13.1
Flow chart for construction and screening of somatic cell hybrids containing chromosome 18.

A.   Overlay Fusion

The fusion of two cell lines can be performed in the presence of a fusing agent. The most commonly used fusing agent is polyethylene glycol (PEG). Although PEG aids in fusing the membranes of two cells, it is also toxic to the cells. The optimum conditions for obtaining fusion with minimal cell death due to PEG toxicity, both in terms of the percentage of the PEG solution used as well as the time for which the cells are exposed to the PEG, must be empirically determined; however, a 1-min exposure to a 50% PEG solution is usually a good starting condition.

Day 1

1. Seed ten 60-mm plates with 6×10^5 cells of UCW206.

2. Incubate overnight at the nonrestrictive temperature of 34°C; by the next day, the plates should be 60 to 80% confluent.

Day 2

1. Centrifuge 2×10^7 EBV-transformed lymphoblasts derived from a patient with a chromosome-18 deletion at 1.5 K for 5 min.

2. Remove media and resuspend cell pellet in 10 ml serum-free DMEM. Pellet cells at 1.5 K for 5 min.

3. Repeat serum-free wash and centrifugation two more times.

4. Resuspend the cell pellet in 20 ml serum-free DMEM containing 20 µl PHA-P (Gibco/BRL).

5. Wash UCW206 cells three times in serum-free DMEM.

6. Remove last wash from UCW206 cells and overlay each plate with 2 ml washed lymphoblast cells. Make sure that the cells are evenly distributed over each plate.

7. Return UCW206 cells to the incubator for 30 min or until the lymphoblast cells have settled onto the UCW206 cells.

8. Remove two plates from the incubator and carefully aspirate off the media.

9. Add 2 ml 50% PEG (Gibco/BRL) to each plate. Tilt plates to distribute PEG evenly over the surface of the cells.

10. After 1 min, add 2 ml DMEM + 10% FCS and swirl to dilute PEG.

11. Remove PEG/media and wash twice more with DMEM + 10% FCS to remove remaining PEG from the cells.

12. Overlay the cells with 4 ml DMEM + 10% FCS and incubate overnight at 34°C.

Day 3

1. Remove plates from the incubator and swirl plates to dislodge any remaining lymphoblast cells from the attached fibroblasts.

2. Remove media and replace with fresh DMEM + 10% FCS.

3. Return plates to an incubator that is set at the restrictive temperature of 39°C.

4. Change media every 4 to 5 days.

Days 21 to 28

1. Between days 21 and 28, colonies of cells should be visible on the plate (from 1 to 10 colonies per plate).

2. On the bottom of each plate, circle the location of growing colonies.

3. Remove media from the plate and place a sterile cloning ring around each colony.

4. Fill the cloning ring with trypsin and let sit for 5 min.

5. Gently triturate the trypsin with a pasteur pipet to dislodge the colony, and transfer the cells to a 24-well plate.

B.   Screening Somatic Cell Hybrid Clones

The lymphoblast cell line used in any fusion has two chromosome-18 homologues; therefore, the somatic cell hybrids that survive the selection can retain either or both chromosome-18 homologues. Since only cell hybrids that have retained the deleted chromosome-18 homologue are of interest, further characterization of the somatic cell hybrids is necessary to eliminate hybrids that have retained a normal chromosome-18 homologue. Furthermore, exposure to PEG and fusion can lead to nonspecific chromosome breakage; therefore, it is important to ensure that the chromosome-18 homologue present in the somatic cell hybrid represents the originally reported chromosome aberration. Traditionally, each somatic cell hybrid was characterized using trypsin-Giemsa staining; however, additional methods have been developed that speed the analysis.

1.   PCR

PCR analysis of the resulting somatic cell hybrid clones represents a rapid method for identifying clones that appear to have retained the deleted chromosome-18 homologue.[4] Sequence-tagged sites (STSs) that map to different regions of the chromosome, both within and outside the putative deletion, are used to screen the hybrid clones.[4-6] A somatic cell hybrid that has retained a normal chromosome 18 will amplify all the chromosome-18-specific STSs, while a somatic cell hybrid retaining a deleted chromosome-18 homologue will amplify only STSs that map outside the deleted region. This screening can be performed several days after the somatic cell hybrids are placed in the 24-well plates.

1. A hybrid clone that is 50 to 100% confluent is trypsinized using standard conditions.

2. One third of the cell suspension is transferred to a new 24-well plate with fresh media, and two thirds of the cells are transferred to an Eppendorf tube.

3. The cells are pelleted for 1 min in an Eppendorf microfuge. The supernatant is removed, and the cells are resuspended in 25 to 200 µl polymerase chain reaction (PCR) buffer + proteinase K (10 mM Tris-HCl, pH 8.3; 50 mM KCl; 0.01% w/v gelatin; 100 µg/ml proteinase K), depending on the size on the size of the cell pellet.

4. The cell suspension is incubated at 60°C for 1 hr followed by a 15-min incubation at 100°C.

5. PCR is performed in a final volume of 25 µl with 2.5-µl DNA sample, 1 µM each primer, 250 mM dNTP, 2% formamide, 0.6 U AmpliTaq DNA polymerase in buffer containing 10 mM Tris-HCl (pH 8.3), 50 mM KCl, 0.01% w/v gelatin, and the appropriate MgCl$_2$ concentration (1.0 to 3.0 mM) for each set of primers. Amplification is performed with an initial denaturation step at 96°C for 90 sec and 35 cycles of denaturation at 96°C for 30 sec and annealing/extension at the appropriate temperature for the primers (50

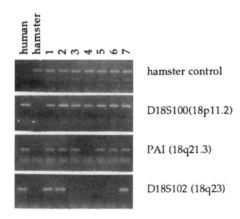

FIGURE 13.2
PCR screening of chromosome-18-specific somatic cell hybrids. Cells were harvested and treated as stated. The marker and its location on chromosome 18 are shown.

to 65°C) for 30 sec with a final extension step at 72°C for 7 min in a thermocycler. The amplified products are normally separated on a 1.5% agarose gel.

Figure 13.2 demonstrates typical results that might be achieved with PCR screening from a fusion experiment using a lymphoblast cell line containing a del(18)(q23). A hamster-specific primer set was used to ensure that each of the hybrid DNA samples were subject to PCR. (Residual media in a cell pellet can inhibit PCR amplification.) Three STSs that map along the chromosome were tested, one mapping to 18p11, one to 18q21.3, and one to 18q23. Hybrids 1, 2, and 7 amplified all three STSs, demonstrating that they had retained a normal chromosome 18. Hybrids 3, 5, and 6 amplified the two proximal STSs but failed to amplify the STS mapping to 18q23, suggesting that these cell hybrids had retained the del(18q23). Hybrid 4 amplified the 18p11 primer but failed to amplify the 18q21.3 or 18q23 primers. This pattern of amplification is inconsistent with retention of either the normal homologue or the deleted homologue and likely represents a chromosome 18 that was broken during the fusion process. From PCR experiment, four of the seven hybrid clones could be eliminated immediately from further analysis.

2.   Fluorescent *in situ* hybridization

Although the PCR screening can provide strong evidence that the appropriate deleted chromosome has been retained, additional screening is important because, as stated above, the fusion process can result in chromosomal breakage, and a deleted chromosome 18 may be retained that is not reflective of the original deleted chromosome. Hybrid 4 in Figure 13.2 represents such an event, and because the break mapped far from the expected breakpoint, such a cell hybrid can be easily identified. Because spurious breakage can be common (the level of breakage is different among various rodent auxotrophic cell lines), it is advisable to identify at least two independent cell hybrids that have identical breakpoints to ensure that the correct deleted chromosome is represented in the cell hybrids. Furthermore, although the correct del(18) may be retained, chromosome 18 fragments generated by the fusion process may also be present in the hybrid that is not detected by the PCR analysis. Fluorescent *in situ* hybridization (FISH) in conjunction with trypsin-Giemsa banding reflects a method for verifying the banding pattern of the deleted chromosome 18 as well as detecting any chromosomal fragments that may be present in the cell hybrid that are of chromosome-18 origin.

1. Metaphase preparations of the somatic cell hybrid cell are harvested using standard conditions.

2. Trypsin-Giemsa banding is performed using standard cytogenetic protocols.

3. Photographs and slide coordinates are taken.

4. The slides are destained by washing the slides twice for 2 min in 70% ethanol, 2 min in 1% HCl in 100% ethanol, and 2 min in methanol.

5. The slides are air dried.

6. The slides are denatured for fluorescent *in situ* hybridization using standard conditions.

7. Chromosome-18-specific painting probes (e.g., Oncor) are used according to manufacturers' protocols.

8. After antibody detection, the identical spreads are observed under the fluorescent microscope.

9. Fluorescent hybridization of the del(18) should be observed. Additional hybridization of any chromosome-18-specific fragments will be easily detectable.

The use of the PCR approach will rapidly eliminate a majority of the somatic cell hybrids that do not contain the chromosome 18 of interest. The fluorescent *in situ* hybridization in conjunction with the traditional trypsin-Giemsa staining will ensure that the somatic cell hybrid chosen is a viable representation of only the chromosomal material represented by the del(18) homologue.

III.   Developing a Somatic Cell Hybrid Mapping Panel

The development of a somatic cell hybrid mapping panel involves the acquisition of several patient samples with deletion breakpoints that are located at different points along the chromosome. The resolution of the mapping panel will be directly proportional to the number of samples that have been obtained, the variation in the location of the breakpoints, and the identification of markers that determine the order of the chromosome breakpoints. Figure 13.3 illustrates a set of somatic cell hybrids that have been derived from patients with various deletions and translocation associated with chromosome 18.

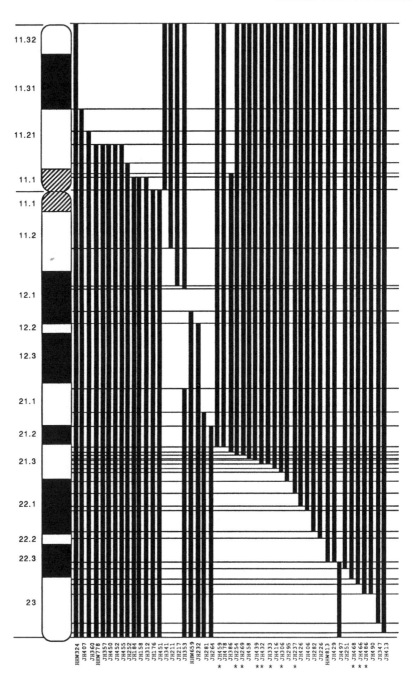

FIGURE 13.3

A deletion mapping panel for chromosome 18. An ideogram of chromosome 18 is shown on the left. The
black bars represent the amount of chromosome-18 material that is present in each cell line. The asterisks
represent somatic cell hybrid cell lines derived from patients with the typical 18q- syndrome phenotype.

Based on the ability to select a deleted chromosome 18, the asparaginyl tRNA synthetase maps to 18q21.2; however, if a deleted chromosome 18 is lacking this 18q21.1 region, it will not be possible to select for a cell hybrid that has a deletion of that region. There are two approaches for obtaining somatic cell hybrids that have such a deletion. In the first method, a different auxotrophic rodent parent cell line that selects for the same chromosome can be used. In the case of chromosome 18, the thymidylate synthetase selectable marker maps to 18p; therefore, the rodent cell line with a mutation in this gene will complement the asparaginyl tRNA synthetase cell line in its ability to select for chromosome 18q21.1 deletions. However, having an alternative rodent cell line for somatic cell hybrid selection is not common. The alternative method is to fuse the lymphoblast cells with a rodent cell line that selects for another chromosome and to screen the resulting somatic cell hybrid colonies for the nonselective retention of the deleted chromosome 18. This approach was used to isolate somatic cell hybrids JH341, JH211, and JH217 that only have 18p material, as shown in Figure 13.3.

If the patients from which a set of somatic cell hybrids have been derived have a chromosomal syndrome, the location of the critical region for a specific syndrome can be determined. As shown in Figure 13.3 the cell lines with asterisks are derived from patients with the classic 18q- syndrome.[7-9] With this set of hybrids, it has been possible to determine that the most terminal band of 18q23 is the critical chromosomal region involved in the clinical features associated with the syndrome, since this is the chromosomal region that is deleted in all of the patients. Therefore, through the isolation of somatic cell hybrids and comparison of the deletions, information of clinical importance can be provided, as well as mapping resources for other laboratories.

IV. Mapping Disease Genes Using a Somatic Cell Hybrid Mapping Panel

One method for identifying candidate disease genes is to determine a gene's location relative to the location of the region known to harbor a disease gene. With respect to a chromosomal deletion syndrome, placing a gene within the critical chromosomal region that is deleted in all patients with the disorder will identify it as a gene that may be involved in the etiology of the syndrome. With respect to the 18q- syndrome, any gene involved in the syndrome must map within 18q23, the smallest region that is deleted in all patients with the syndrome. Several genes have been mapped to the distal half of the chromosome 18 and include plasminogen activator inhibitor (PAI),[5] gastrin-releasing peptide (GRP),[5] and nuclear factor of activated T cells (NFATc).[10] Figure 13.4 shows the mapping of several of these genes relative to chromosomal breakpoints in patients with the classic 18q- syndrome phenotype. Both GRP and PAI map within 18q21.3 and therefore map proximal to the 18q- critical region. In contrast, NFATc is deleted in all of the patients, localizing it to the 18q- critical

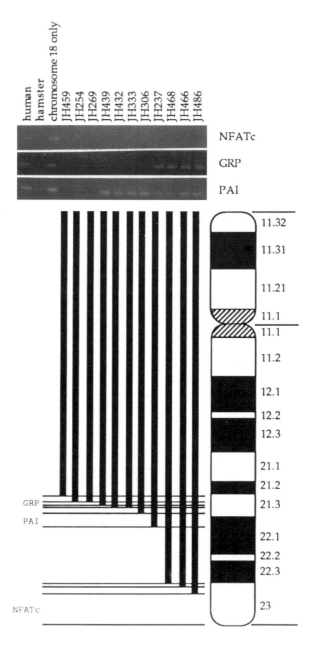

FIGURE 13.4
Mapping of disease genes relative to chromosomal breakpoints in somatic cell hybrids. Primers specific for each of the genes were used in PCR reactions with the input genomic DNA as shown.

region. NFATc is a part of a transcription factor complex that activates cytokines and is involved in the immune response. Because IgA deficiency as well as autoimmune disease are clinical features observed in a subset of patients with the 18q- syndrome,[11,12] it is possible that hemizygosity of NFATc may be involved in the occurrence of IgA deficiency in this syndrome. Therefore, by further mapping known genes relative to deletions in patients with specific clinical phenotypes, it is possible to identify candidate genes involved in a specific clinical symptom.

V. Conclusions

The construction of somatic cell hybrid deletion mapping panels represents a rapid method for mapping genes on a specific chromosome. The advantage of this mapping over the use of FISH is that gene segments in the STS format can be easily mapped. In contrast, even full-length cDNAs may be difficult to localize by FISH unless a CCD camera is attached to the fluorescent microscope. Furthermore, whereas the location of a gene can be rapidly determined relative to a number of patients with an identical disorder (as in the case of a deletion syndrome) when somatic cell hybrids have been derived from the patients, such comparison studies would be time consuming at best using a fluorescent approach. Therefore, somatic cell hybrid mapping panels are of multiple uses: rapid mapping of PCR- or non-PCR-based genes and the generation of phenotype-genotype maps in the characterization of deletion syndromes.

References

1. Patterson, D., *Somat. Cell Mol. Genet.,* 13, 365, 1987.
2. Cirullo, R. E., Arredondo-Vega, F. X., Smith, M., and Wasmuth, J. J., *Somat. Cell Mol. Genet.,* 9, 215, 1983.
3. Nussbaum, R. L., Walmsley, R. M., Lesko, J. G., Airhart, S. D., and Ledbetter, D. H., *Am. J. Hum. Genet.,* 37, 1192, 1985.
4. Kline, A. D., Rojas, K., Mewar, R., Moshinsky, D., and Overhauser, J., *Genomics,* 13, 1, 1992.
5. Overhauser, J., Mewar, R., Rojas, K., Lia, K., Kline, A. D., and Silverman, G. A., *Genomics,* 15, 387, 1993.
6. Rojas, K., Straub, R. E., Kurtz, A., Feder, M., Mewar, R., Gilliam, T. C., and Overhauser, J., *Genomics,* 14, 1095, 1992.
7. Kline, A. D., White, M. E., Wapner, R., Rojas, K., Biesecker, L. G., Kamholz, J., Zackai, E. H., Muencke, M., Scott, Jr., G. I., and Overhauser, J., *Am. J. Hum. Genet.,* 52, 895, 1993.
8. Strathdee, G., Zackai, E. H., Shapiro, R., Kamholz, J., and Overhauser, J., *Am. J. Med. Genet.,* 59, 476, 1995.
9. Strathdee, G., Sutherland, R., Johnsson, J. J., Sataloff, R., Kohonen-Corish, M., Grady, D., and Overhauser, J., *Am J. Hum. Genet.,* 60, 860, 1997.

10. Li, X., Ho, S. N., Luna, J., Giacalone, J., Thomas, D. J., Timmerman, L. A., Crabtree, G. R., and Francke, U., *Cytogenet. Cell Genet.,* 68, 185, 1995.

11. Hecht, F., *Lancet,* 1, 100, 1969.

12. Wilson, M. G., Towner, J. W., Forsman, I., and Sims, E., *Am. J. Med. Genet.,* 3, 155, 1979.

Chapter

14

Gene Mapping Using Somatic Cell Hybrids

Nigel K. Spurr, Mark Bouzyk,
and David P. Kelsell

Contents

I. Introduction

Human-rodent somatic cell hybrids have proven to be a useful tool for gene mapping. Initially, they were used to map expressed gene products — for example, enzymes or cell surface receptors. This chapter details methods for DNA-based gene mapping techniques using somatic cell hybrids.

A. Somatic Cell Hybrids

The first somatic cell hybrids were produced more than 30 years ago from mixed cultures of two related, but different mouse cell lines.[1] These hybrids arose spontaneously and were isolated because the hybrid cells outgrew the parental populations; however, not all hybrid combinations show this "hybrid vigor". Since the frequency of hybrid formation was often very low, other selective systems were developed.

In 1964, Littlefield[2] showed that the HAT (hypoxanthine-aminopterin-thymidine) selective system[3] could be used to isolate two mutant cell lines: A3-1, an L-cell line that was deficient in hypoxanthine guanine phosphoribosyl transferase (HPRT), and B34, a cell line deficient in thymidine kinase (TK). When these lines were cultured in the presence of HAT medium, they died. The aminopterin (nowadays substituted by methotrexate) inhibits dihydrofolate reductase and prevents *de novo* synthesis of purines and one-carbon transfer reactions, including the conversion of deoxyuridylic acid (1 UMP) to thymidylic acid (TMP). Unless cells exposed to HAT were able to utilize exogenous pyrimidines and purines (for example, thymidine and hypoxanthine), they would not survive. Only spontaneous fusion products grew, since the hybrids complemented the genetic defects of the parents and contained both HPRT and TK activities.

The first continuous human-mouse somatic cell hybrids produced were from a spontaneous fusion occurring between mouse L cells deficient in TK and normal human embryonic fibroblasts.[4] The hybrids were selected in HAT. The mouse cells were killed, and the hybrids outgrew the human cells. It was found that as the hybrids grew, the mouse chromosomes were all retained but the human chromosomes were preferentially lost. This observation provided the key to their use in gene mapping, enabling the assignment of human genes and sequences. As the mouse cell line was thymidine-kinase deficient, the mouse-human hybrids selected in HAT were required to retain the human gene for thymidine kinase. After several generations in culture, the majority of the hybrid cells had retained only one chromosome in common, chromosome 17; therefore, the TK gene could be assigned to chromosome 17.[5,6]

Gene-mapping studies were first carried out using enzyme loci. Enzymes are relatively simple to assay, and human and rodent forms can often be separated electrophoretically into species-specific isoenzymes.[7] The cell surface antigens are another major group of products expressed in somatic cell hybrids. Human mono-clonal antibodies have been used to assign specific cell surface antigens chromo-somally using somatic cell hybrids.[8]

Somatic cell-hybrid gene-mapping techniques dependent on the expression of a gene product are not always possible. This is due to the phenomenon of "extinction" or loss of the expressed phenotype after fusion of two parent cells.

B.   Gene Mapping Using DNA-Based Technology

In the late 1970s, recombinant DNA technology was applied to human genetic analysis. It had a major impact on the ability to assign genes to chromosomes, as there was no longer a requirement for expression of a gene product. This meant that not only was it possible to assign specific genes, but also any randomly isolated piece of DNA that was free of repetitive DNA sequence. As with gene assignments using protein products, those using DNA also depend on the loss or retention of specific chromosomes or chromosome fragments in somatic cell hybrids. The following points should be considered when using somatic cell hybrids for DNA-based gene mapping:

- Somatic cell hybrid cell lines should not be kept in long-term culture; chromosomes not under direct selection will be lost at random.

- The chromosome content of each hybrid line should be monitored using cells or DNA from the same passage as that DNA extracted for use in mapping studies.

- Whether using Southern blotting or polymerase chain reaction (PCR) amplification, a species difference must be determined.

The first chromosome assignments using cloned DNA were of the human alpha- and beta-globin genes. This was achieved using solution hybridization of alpha- and beta-globin cDNAs with genomic DNA prepared from several hybrid cell lines containing different human chromosomes. The alpha-globin locus was assigned to chromosome 16,[9] and the beta-globin gene to chromosome 11.[10] These initial studies led to an explosion of gene assignments using DNA probes on genomic DNA extracted from somatic cell hybrids.

The impact of the use of DNA probes for human gene mapping was first seen at the Sixth Human Gene Mapping Workshop, Oslo (1981). At this meeting, a total of 45 cloned DNA segments had been identified, of which 35 had been assigned. The rate of assignment has increased exponentially following the development of mapping using PCR amplification. More than 5000 genes have now been assigned using panels of human-rodent somatic cell hybrids.[11]

Subchromosomal gene localization is also possible with the use of somatic cell hybrids. Hybrids have been constructed from the fusion of rodent cells to human cells which harbor specific chromosomal deletions or translocations. The chromosomal aberration can be maintained in the somatic cell hybrid, and gene location determined by their presence or absence. An example of regional mapping is shown in Figure 14.3 in Section III.

Somatic cell-hybrid technology can be used to construct high-resolution contiguous maps of mammalian chromosomes by radiation hybrid (RH) mapping.[12] The method involves exposing a human-rodent somatic cell containing a single human chromosome to a high dose of X-rays in order to break the chromosomes into several fragments. The fragment-bearing hybrid cells that are nonviable can be removed by fusing with a normal rodent cell line. Growth in HAT medium selects against the normal nonfused rodent cells, assuming they are deficient in the enzyme HPRT. Hybrid clones obtained from fusion each contain a unique set of fragments from the original human chromosome. Each clone is then screened for the presence or absence of DNA markers assigned to that chromosome under investigation. The main assumption of RH mapping is that the farther apart two loci are on a chromosome the more likely that radiation will break the chromosome between them, thus placing the loci on two separate fragments. By estimating the probability of breakage and also the distance between loci, it is possible to determine their order. RH mapping can integrate both genetic and transcription maps, plus any other sequenced genome landmarks on a single map.

Currently, the use of DNA from somatic cell hybrids forms the core of many mapping strategies, particularly those aimed at the rapid "binning" of genes or DNA fragments to individual chromosomes. This chapter concentrates on the use of hybrids in gene mapping using either Southern blot analysis or PCR-based methods.

II.   Methods

A.   Somatic Cell Hybrids

This section describes the methods used in the selection, maintenance, and characterization of hybrid cell lines.

1.   Cell culture and selection

Methods for the production of somatic cell hybrids and general methods of cell culture are not covered in this chapter. Several excellent books already exist which give precise details on these methodologies; however, it is valuable to highlight some of the problems that must be considered when culturing these cells. The majority of somatic cell hybrids have been prepared by fusing a rodent cell line

TABLE 14.1
Examples of Some of the Commonly Used Methods of Biochemical Selection

Chromosome	Selected Marker	Ref.
1p34.1-p34.3	CTP synthetase (CTPS)	30
5cen-q11	Leucyl tRNA syntetase (LARS)	31
5q11.2-q13.2	Dihydrofolate reductase (DHFR)	32
5q23	Diphtheria toxin receptor (DTS)	33
5q35	Chromate resistance (CHR)	31, 34
9q12-pter	Methylthioadenosine phosphorylase (MTAP)	35
9cen-q24	Adenosine kinase (ADK)	36
16q24.2-qter	Adenine phosphoribosyltransferase (APRT)	37
17q23.3-q25.3	Thymidine kinase 1 (TKI)	38
18p11.31-p11.21	Thymidylate synthase (TYMS)	39
18q21.2-q21.3	Asparaginyl-tRNA synthase (NARS)	40
20q12-q13.11	Adenosine deaminase (ADA)	41
Xq26	Hypoxanthine phosphoribosyltransferase (HPRT)	42

deficient in either the HPRT or TK loci with a human cell line with limited growth potential (for example, peripheral blood lymphocytes). The hybrid cells are then cultured in HAT medium, and in the resultant clones the function of the deficient genes is complemented by the human gene. HPRT maps to the human X chromosome and TK to chromosome 17; therefore, complete, or fragments of these, human chromosomes will be retained in the hybrid cell as long as it is under selection. If the HAT selection is removed, then these chromosomes will segregate and be lost with time.

The remaining human chromosomes in the hybrids which are not under any selective pressure will be lost in an apparently random fashion; however, there is some evidence that these losses are not completely random, and the retention of certain chromosomes may give the cells an added growth advantage. Conversely, the retention of some chromosomes may lead to senescence or reversion to a slower growing phenotype. In our experience, obtaining hybrids that contain human chromosomes 9 and 10 has always been difficult and often requires the use of alternative methods of selection.

Other methods applied to selecting individual chromosomes are based on the presence of specific genes; for example, human chromosome 16 can be selected following fusion of mouse cells deficient in the adenine phosphoribosyltransferase (APRT) gene. Other methods of biochemical selection are based on the use of complementing rodent cells deficient in certain enzymes, particularly those involved in cell metabolism (Table 14.1).

TABLE 14.2
Dominant Selectable Markers

Selection System	Ref.
G418/neomycin phosphotransferase II (neo)	43, 44
Mycophenolic acid/xanthine-guanine phosphoribosyltransferase (XGPRT)	45
Hygromycin B/hygromycin B kinase (hph)	46
Puromycin/puromycin N-acetyl transferase (pac)	47
Histidinol/histidinol dehydrogenase (hisD)	48
Tryptophan/tryptophan synthase (trpB)	48
Phleomycin/ble gene	49, 50
Albizzin/asparagine synthetase	51
Methotrexate/dihydrofolate reductase (DHFR)	52

The above methods of selection are all based on using suitable mutant recipient cells. An alternative to this is the introduction of a marker conferring a dominant selectable phenotype into the human cells and then fusing them with an established rodent cell line. The integration of the dominant marker is usually at random, and the tagged chromosome is isolated by microcell fusion to rodent cells. The most widely employed methods of selection are outlined in Table 14.2

Key points to consider when culturing somatic cell hybrids are

- Keep the cell line in culture for the shortest possible time.
- Prepare sufficient seed stock before commencing large-scale culture.
- Keep the cells under the appropriate selective pressure.
- Keep the cells free of mycoplasma by checking stocks regularly.
- Recharacterize the human content of the hybrids from cells harvested at the same time as those used to prepare DNA.

2. Characterization

As the chromosome content of a set of hybrids will vary after culture, it is essential to check the human chromosome content of each line. Several methods may be used to analyze the human chromosome content of each hybrid.

Marker analysis

The early experiments in somatic cell genetics utilized enzyme selection or isoenzyme analysis to confirm the identity of the human chromosomes retained in the hybrids. This has been largely superceded by DNA analysis using either Southern blotting or PCR amplification. PCR is the method of choice with an extensive range of PCR primer pairs available for all human chromosomes.[13-16]

Cytogenetic analysis

Metaphase chromosome spreads stained with G11, a Giemsa stain, allows the discrimination between human and mouse chromosomes. Mouse cells are stained deep purple or mauve, while the human chromosomes are lightly colored pink.[17] This method quickly determines the number of human chromosomes in a hybrid; however, it is not always possible to identify the chromosome accurately.

The precise identification of the chromosomes in a hybrid requires further banding techniques — for example, quinacrine mustard staining.[18] Recently, these two methods have been dropped in favor of fluorescence *in situ* hybridization (FISH).[19] Numerous chromosome-specific reagents are now available to allow the accurate identification of each chromosome or fragment. These include chromosome-specific paints, centromere-specific probes, and regional probes usually greater than 5 to 10 kb in length — for example, yeast artificial chromosomes (YACs) containing human genomic DNA.

All of these methods allow the identification of a specific chromosome or region. It is difficult, however, to detect small deletions or rearrangements of chromosomal material, and one must always be aware of the possibilities of misassignment of a DNA fragment if only one panel of hybrids is used.

B.   Analysis of Somatic Cell Hybrids for DNA-Based Gene Mapping

This section describes methods for DNA-based gene mapping using Southern blot analysis or PCR.

1.   Extraction of genomic DNA

Approximately 10^8 cells are harvested (usually by scraping), washed in ice-cold phosphate buffered saline (PBS), and then centrifuged in a 50-ml Falcon tube at ~500 g for 5 min. The PBSA is removed, and the resulting cell pellet is used for DNA extraction.

The cell pellet is resuspended in 15 ml of 10 mM Tris-HCl (pH 7.6) and 100 mM ethylenediaminetetra acetate (EDTA), pH 8.0. Sodium dodecyl sulfate (SDS) is added to 0.5% and proteinase K (Boehringer Mannheim) to 0.1 mg/ml before incubation at 50°C for at least 4 hr. After this, NaCl is added to 0.1 M and the solution then extracted with phenol. The solution is extracted a second time with a 1:1 mixture of phenol:chloroform.

Finally, the DNA is precipitated by the addition of 2 volumes of ethanol and spooled out using a hooked glass rod. The DNA is dried briefly under vacuum and resuspended in 0.5 to 1 ml of TE (pH 7.6), depending upon the yield. This solution is then left overnight to allow the DNA to dissolve completely.

The DNA is then used for gene mapping using Southern blot analysis, PCR, or a combination of both techniques. These methods are described below.

2.   Southern blot analysis

Homologous sequences in genomic DNA can be detected by hybridization of a labeled probe to DNA fragments generated by restriction endonuclease digestion, separated on an agarose gel, and transferred to a solid support. This technique was first described by Southern in 1975.[20] This method has been used to map many hundreds of genes/DNA fragments and is dependent on the identification of a species difference between the human and rodents species. With probes prepared from random fragments of genomic DNA, there is often little or no highly homologous sequences in rodents, so the problem of cross-hybridization is reduced. However, cross hybridization can be a problem when conserved sequences such as cDNAs are used as probes. If fragments hybridize to both species, the testing of several different restriction enzymes may be required to distinguish clearly the species-specific fragments.

Restriction enzyme digestion of genomic DNA

Restriction enzymes can be purchased from a number of manufacturers, including Gibco BRL, Boehringer Mannheim, New England Biolabs, and NBL. Digestions are carried out using the buffers supplied by the manufacturers and according to their recommendations.

1.   10 μg of genomic DNA are generally digested for 3 hr using 30 U of restriction endonuclease incubated at the optimum temperature for activity of the endonuclease used.

2.   After digestion, the DNA fragments can be separated by electrophoresis using submarine horizontal agarose gels in 1× TAE buffer, run at 1 to 2 v/cm. 1% agarose gels made up in 1× TAE buffer (0.04 M Tris-acetate; 0.001 M EDTA, pH 8.0) are suitable for most applications.

3.   0.1 μg/ml of ethidium bromide is added to the running buffer which enables the DNA fragments to be visualized using an ultraviolet transilluminator. DNA fragment lengths can be determined by using DNA standards of known size (for example, Gibco BRL 1-kb ladder and *Hind*III-digested λ DNA). Permanent records of the gel can be obtained by photography through an orange filter using, for example, a Polaroid Land camera and Type 57 film.

1× TAE buffer

40 mM Tris-acetate (pH 7.8)
2 mM EDTA (pH 8.0)

Transfer of DNA from agarose gel

After electrophoresis, the DNA is transferred to, for example, a Hybond N+ nylon membrane (Amersham International). The gel is soaked in 0.4 *M* NaOH for 20 min with gentle agitation and then blotted in the same concentration of NaOH. The denatured DNA is transferred to the membrane by capillary transfer in the NaOH overnight. After blotting, the filter is washed in 5× saturated sodium citrate (SSC). Hybond N+ filters do not require fixing, as the membrane is positively charged, electrostatically binding the DNA upon transfer.

Generation of labeled DNA probes

The randomly primed labeling technique is used to generate radioactively labeled DNA fragments for use as probes in Southern blot analyses;[21] 50 to 100 ng of probe are added to the labeling reaction (for example, the Amersham Multiprime Kit), according to the manufacturer's protocol.

After labeling, the unincorporated dNTPs are removed by running the reaction mixture through a Sephadex G50 column (Pharmacia) equilibrated in 3× SSC. Typical incorporation of labeled dCTP using this technique is usually greater than 80%, with a specific activity of greater than 10^9 cpm/mg of DNA.

Hybridization of labeled DNA probes to DNA immobilized on nylon filters

Hybridization is carried out at 65°C in glass tubes placed in a commercial hybridization oven (for example, Hybaid, Ltd.). Filters are prehybridized for 2 to 4 hr in varying amounts of prehybridization solution, depending upon filter size and number of filters. The prehybridization mix is then removed and the hybridization mix containing the denatured probe added. The radioactively labeled probe is denatured by boiling for 5 min prior to adding to the hybridization mix. Hybridization is generally carried out overnight.

Prehybridization solution

5× SSC
5× Denhardt's solution
40 m*M* NaPO$_4$ (pH 6.5)
0.1 mg/ml denatured sheared herring sperm DNA (Sigma)

Hybridization solution

Prehybridization solution
10% (w/v) dextran sulphate
Denatured probe

100× Denhardt's Solution

2% (w/v) bovine serum albumen (BSA) (Fraction V, Sigma)
2% (w/v) Ficoll 400 (Sigma)
2% (w/v) polyvinylpyrrolidone 360 (Sigma)

Washing filters

When the hybridization is complete, the hybridization mix is removed and the filters washed. Three quick washes are performed at room temperature, followed by one 10-min wash at 65°C, using Wash 1.

Wash 1

0.2× SSC
0.1% (w/v) SDS

Autoradiography

The filter is wrapped in plastic film (for example, Saran wrap) and placed in an X-ray film cassette. We routinely use Kodak X-OMAT AR film and expose the filter to the film, usually at –70°C for 1 to 5 days to visualize the hybridized fragments. The film is developed using a commercial film developing system (for example, Fuji, Ltd.).

C.   Analysis of Hybrids Using Polymerase Chain Reaction Amplification

1.   DNA amplification by polymerase chain reaction

The polymerase chain reaction (PCR) is used to amplify DNA between two regions of known sequence.[22] Oligonucleotide primers are synthesized using an automated DNA synthesizer (Model 380B, Applied Biosystems), working on cyanoethyl phosphoramidite chemistry. A general guide for primer design is that each primer is about 50% GC-rich, not self complementary, and not complementary to each other. Ideally, the oligonucleotides should be between 17 and 24 bp long with a melting temperature of about 60°C (assuming A, T = 2°C; G, C = 4°C). Other important variables to consider in obtaining a PCR product are annealing temperature, $MgCl_2$ concentration, and number of cycles.

As well as following the primer design conditions as described above, other points should be considered as to which region of a gene should be targeted to generate gene-specific primers, some of which are listed below:

1.   Intronic sequence: reduced sequence conservation between gene family members and across species

2. 3′ untranslated region: as above

3. Sequences aligned between gene family members or across species to identify regions with a number of nucleotide mismatches

PCR is performed in 50-µl reactions containing 30 ng template DNA, 1× PCR buffer (Boehringer Mannheim), 20 pmol primer, 200 mM dNTPs (Boehringer Mannheim), and 1 U of Taq polymerase (Boehringer Mannheim) on a GeneAmp 9600 thermal cycler (Perkin-Elmer Cetus). Reaction mixtures are given 30 cycles of 30 sec at 94°C, 30 sec at 55°C, and 30 sec at 72°C. For larger products, the extension time may have to be lengthened. To generate human-specific PCR products, the annealing temperature may have to be increased.

Then, 15 µl of each PCR reaction are electrophoresed through submerged horizontal gels (Gibco BRL and Scotlab) in 1× TBE buffer (0.5× 0.045 M Tris-borate; 0.001 M EDTA, pH 8.0) run at 3 to 5 v/cm. 2% agarose gels made up in 1× TBE buffer are used for most applications to achieve separation of PCR fragment(s), usually in the range of 100 to 1000 bp. The running buffer contains 0.1 µg/ml ethidium bromide, which enables the DNA fragments to be visualized using a UV transilluminator. DNA fragment lengths are determined by using DNA standards of known size. Permanent records of the gels may be obtained by photography through an orange filter using, for example, a Polaroid Land camera and Type 57 film.

1× TBE

   45 mM Tris-borate
   1 mM EDTA

III. Results

In this section, we describe examples of the assignment of a number of genes/random DNA sequences using the DNA mapping strategies and analyses as described in Section II.

A. Southern Blot Mapping: Assignment of the Genomic Clone p3.1 to Chromosome 5[23]

p3.1 is a 5.8-kb genomic clone isolated from a lambda Charon 4A library containing human DNA digested with the restriction endonuclease *Eco*R1. It does not contain repetitive DNA sequences and detects a relatively frequent two-allele restriction fragment length polymorphism (RFLP) at a *Msp*1 restriction site. To localize this polymorphic DNA fragment, Southern blot hybridization to a panel of complex human-rodent somatic cell hybrids (and species controls) was performed.

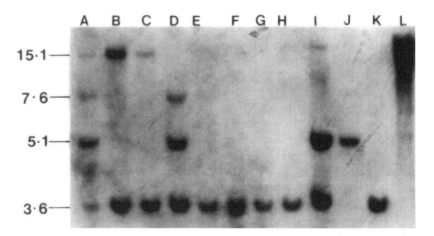

FIGURE 14.1

Southern blot of various genomic DNA samples digested with *Eco*R1 and hybridized with radiolabeled p3.1. The autoradiograph shows human (track J), mouse (track K), and three positive hybrids: CTP34B4 (track A), Dur4R3 (track D), and Mog 34A4 (track I). Six negative hybrids are shown: CTP41A2 (track B), 3W4CL5 (track C), Sir74ii (track E), DT1.2.4 (track F), DT1.2 (track G), and Horp9.5 (track H). A detailed description of the chromosomal content of each hybrid can be found in Table 14.3. (The fragments at 15.1 and 7.6 kb are partials due to incomplete digestion of DNA.)

An example of a Southern blot of genomic DNA from a hybrid mapping panel digested with the enzyme *Eco*R1 hybridized to p3.1 is shown in Figure 14.1. p3.1 hybridizes to a single band in human DNA of 5.0 kb (track J). The fragments at 15.1 and 7.6 kb are partials due to incomplete digestion of DNA. A species difference was observed, as in mouse DNA p3.1 hybridizes to a smaller, single band of 3.6 kb (track K). The human-specific 5.0-kb fragment was present in a number of hybrids — for example, CTP34B4 (track A), Dur4R3 (track D), and MOG34A4 (track I). Six negative hybrids are shown: CTP41A2 (track B), 3W4CL5 (track C), Sir74ii (track E), DT1.2.4 (track F), DTI.2 (track G), and Horp9.5 (track H). The results from all hybrids analyzed were consistent with the presence of human chromosome 5, the remaining chromosomes being discordant (Table 14.3). Therefore, p3.1 can be assigned to chromosome 5.

B.   PCR Mapping

1.   Monochromosomal panel: chromosomal assignment of novel expressed sequence tagged sites (eSTS)[24]

The assignment of transcribed genes identified through partial cDNA sequencing to a specific chromosome has been carried out for more than 4000 human gene transcripts.[25-27] A human-rodent monochromosomal somatic cell hybrid panel (Table 14.4) was used to assign four eSTSs.

TABLE 14.3
Assignment of Clone 3.1

Hybrid Name	1	2	3	4	5	6	7	8	9	10	11	12	13	14	15	16	17	18	19	20	21	22	X
MOG34A4	+	−	+	+	+	+	+	+	−	+	+	+	+	+	−	+	−	+	+	−	+	−	+
CTP34B4	+	+	+	−	+	+	+	+	−	−	−	+	−	+	−	+	+	+	−	−	−	−	+
DUR4R3	−	−	+	−	+	−	−	+	−	−	+	+	+	+	t	−	+	+	−	+	+	−	−
FG10	−	+	−	−	+	−	−	+	−	+	−	−	−	−	+	−	−	+	−	−	+	−	+
DELTSCL16	−	−	−	−	+	−	−	−	−	−	−	−	−	−	−	−	−	+	−	−	−	−	−
SIR19A	+	−	+	−	+	−	+	+	−	+	+	+	+	+	+	−	+	+	+	−	+	t	+
CTP41A2	−	+	+	−	−	+	+	−	−	−	−	−	−	−	+	−	−	+	−	−	−	−	+
3W4CL5	−	−	−	−	−	−	+	−	−	+	+	+	−	+	+	−	+	−	−	−	+	−	+
THYB1.3	−	−	−	−	−	−	−	−	−	−	−	−	−	−	−	−	−	−	−	−	+	−	+
HORP9.5	−	−	−	−	−	−	−	−	+	+	+	−	+	−	−	−	−	−	−	−	−	+	+
SIF15	−	+	−	−	−	+	+	−	−	+	−	−	−	+	+	−	−	−	−	+	−	−	+
SIR74ii	+	+	+	+	−	−	−	−	−	−	−	−	+	+	+	−	+	t	−	−	+	t	+
DTI.2.4	−	−	+	−	−	−	−	−	−	−	+	−	−	−	+	−	+	+	−	+	+	−	+
DTI.2	−	−	+	−	−	−	−	−	−	+	+	−	+	−	+	−	+	+	−	+	+	+	+
FIR5	−	−	−	−	−	−	+	−	−	−	−	−	−	+	−	−	−	+	−	−	−	−	+
FST9/10	−	−	+	+	−	+	−	+	−	+	−	+	+	−	+	−	−	+	−	+	−	+	+
298-20b	+	+	−	+	−	−		−	+	−	−	−				−	+	+					+
RJD1	−	−	+	−	−	−	t	−	−	+	−	−	−	−	−	−	−	+	t	t	t	t	+

Note: + signifies presence; − signifies absence; t = trace.

Four unique cDNA sequences with no nucleic acid or protein homology were selected from a human tissue cDNA database held at the U.K. Human Genome Project Resource Centre (Hinxton Hall, Cambridge). Oligonucleotide primer pairs were designed for each of the cDNA fragments using the criteria described in Section II. These oligonucleotides were used in PCR amplification of the panel of monochromosomal somatic hybrid cells to assign cDNAs: AAACQHM, AAACQLS, AAAHOG, and AAAURE to chromosomes 1, 3, 16, and 20, respectively (Figure 14.2). Total human genomic DNA acted as a positive control. Mouse and hamster genomic DNAs were used as negative controls. Primer sequences and product sizes for each cDNA fragment are listed in Table 14.5.

2. Translocation hybrid mapping: assignment of p21 to the chromosomal region 6p21[24]

p21 (Cip1/WAF1) codes for a protein that forms a multiple quaternary complex with a cyclin-dependent kinase, a cyclin, and proliferating cell nuclear antigen.[28] From the published human cDNA sequence,[28] oligonucleotide sequences were designed

TABLE 14.4
Monochromosomal Somatic Cell Hybrid Panel

Hybrid[a]	1	2	3	4	5	6	7	8	9	10	11	12	13	14	15	16	17	18	19	20	21	22	X	Y
GM07299	+	–	–	–	–	–	–	–	–	–	–	–	–	–	–	–	–	–	–	–	–	–	+	–
GM10826B	–	+	–	–	–	–	–	–	–	–	–	–	–	–	–	–	–	–	–	–	–	–	–	–
GM10253	–	–	+	–	–	–	–	–	–	–	–	–	–	–	–	–	–	–	–	–	–	–	–	–
HHW416	–	–	–	+	–	–	–	–	–	–	–	–	–	–	–	–	–	–	–	–	–	–	–	–
GM10114	–	–	–	–	+	–	–	–	–	–	–	–	–	–	–	–	–	–	–	–	–	–	–	–
MCP6BRA	–	–	–	–	/	–	–	–	–	–	–	–	–	–	–	–	–	–	–	–	–	–	/	–
CLNE21E	–	–	–	–	–	+	–	–	–	–	–	–	–	–	–	–	–	–	–	–	–	–	–	–
C4A	–	–	–	–	–	–	+	–	–	–	–	–	–	–	–	–	–	–	–	–	–	–	–	–
GM10611	–	–	–	–	–	–	–	–	+	–	–	–	–	–	–	–	–	–	–	–	–	–	–	–
762-8a	–	–	–	–	–	–	–	–	–	+	–	–	–	–	–	–	–	–	–	–	–	–	–	+
JICL4	–	–	–	–	–	–	–	–	–	–	+	–	–	–	–	–	–	–	–	–	–	–	–	–
1aA9602+	–	–	–	–	–	–	–	–	–	–	–	+	–	–	–	–	–	–	–	–	+	–	+	–
289	–	–	–	–	–	–	–	/	–	–	/	/	+	–	–	–	–	–	–	–	–	–	–	–
GM10479	–	–	–	–	–	–	–	–	–	–	–	–	–	+	–	/	–	–	–	–	–	–	–	–
HORLI	–	–	–	–	–	–	–	/	–	–	–	–	–	–	+	–	–	–	–	–	–	–	/	–
286OH7	–	–	–	–	–	–	–	–	–	–	–	–	–	–	–	+	–	–	–	–	–	–	–	–
PCTBA1.8	–	–	–	–	–	–	–	–	–	–	–	–	–	–	–	–	+	–	–	–	–	–	–	–
DL18TS	–	–	–	–	–	–	–	–	–	–	–	–	–	–	–	–	–	+	–	–	–	–	–	–
GM10612	–	–	–	–	–	–	–	–	–	–	–	–	–	–	–	–	–	–	+	–	–	–	–	–
GM10478	–	–	–	/	–	–	–	/	–	–	–	–	–	–	–	–	–	–	–	+	–	/	+	–
THYB1.3	–	–	–	–	–	–	–	–	–	–	–	–	–	–	–	–	–	–	–	+	–	+	+	–
PGME25NU	–	–	–	–	–	–	–	–	–	–	–	–	–	–	–	–	–	–	–	–	–	+	/	–
HORL9X	–	–	–	–	–	–	–	–	–	–	–	–	–	–	–	–	–	–	–	–	–	–	+	–
853	–	–	–	–	–	–	–	–	–	–	–	–	–	–	–	–	–	–	–	–	–	–	–	+

Note: + indicates presence; – indicates absence; / indicates chromosome translocation, extra chromosomes, or other modifications.

[a]   GMO7299   1 and x
   MCP6BRA   Xqter-Xq13:6p21-6qter
   GM10611   9p stronger than 9q; only 10% of cells with whole 9
   762-8a   10 and Y
   1aA9602+   12, 21, X
   289   13 plus fragments of 8, 11, 12
   GM10479   14, plus part of 16 (probably 16p13.1-16q22.1)
   HORLI   15, liq, part of Xp and proximal Xq
   GM10478   20, 4 (part), 8(part), 22q, X
   PGME25NU   22, part of Xp

following the criteria for primer design as described in Section II. They were primer A (5′-TGTACCCTTGTGCCTCGCTC-3′ at nucleotide position 425–444) and primer B (5′-GGGCGGATTAGGGCTTCCTC-3′ at nucleotide position 578–559). Primer B was

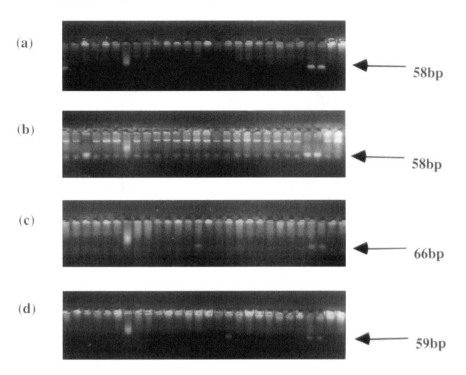

FIGURE 14.2

Chromosomal assignment of eSTS: In all four examples, the order of somatic cell hybrids is as follows: GM07299 (track 1), GM10826B (track 2), GM10253 (track 3), GM10114 (track 4), MCP6BRA (track 5), CLONE21E (track 6), C4A (track 7), GM10611 (track 8), 762-8a (track 9), 1aA9602+ (track 10), 289 (track 11), GM10479 (track 12), HORLI (track 13), 286H7 (track 14), PCTBA1.8 (track 15), GM10612 (track 16), GM10478 (track 17), THYB1.3 (track 18), PgME25NU (track 19), HORL9X (track 20), 853 (track 21), HHW416 (track 22), JICL4 (track 23), and DL18TS (track 24). A detailed description of the chromosomal content of each hybrid can be found in Table 14.4. Additionally, in all cases genomic DNA controls are human (tracks 25 and 26), hamster (track 27), and mouse (track 28). **(a)** PCR amplification of AAACQHM–58-bp PCR product was specifically amplified in the human control (tracks 25 and 26) and in the somatic cell hybrid GM07299 (track 1) which corresponds to human chromosome 1. **(b)** PCR amplification of AAACQLS–58-bp PCR product was specifically amplified in the human control (tracks 25 and 26) and in the somatic cell hybrid GM10253 (track 3), which corresponds to human chromosome 3. **(c)** PCR amplification of AAAAHOG–66-bp PCR product was specifically amplified in the human control (tracks 25 and 26) and in the somatic cell hybrid 2860H7 (track 14) which corresponds to human chromosome 16. **(d)** PCR amplification of AAAAURE–59-bp PCR product was specifically amplified in the human control (tracks 25 and 26) and in the somatic cell hybrid GM10478 (track 17) which corresponds to human chromosome 20.

selected from the 3′ untranslated region of the gene, where there is unlikely to be strong interspecies sequence conservation, so the primers are unlikely to PCR-amplify rodent p21. The expected PCR product size from the cDNA sequence was 154 bp and was confirmed from PCR amplification of a number of cDNA libraries (Figure 14.3a, tracks 2, 5, 6); however, when human genomic DNA was the

TABLE 14.5
ESTS Chromosomal Assignment

cDNA/ HGMP I.D.	Primer Sequence 5'-3'	Chromosome	Product Size (bp)
AAACQHM	AGAACGACATGTGAGAAAAC AATAACATGGTGGGTTGACA	1	58
AAACQLS	TCTGACTGTGGAATAGCATG TCATCTCACTGTGCCTGTG	3	58
AAAAHOG	CTCATACTCAGTTTTTTATGTA AATTAGTGTGGAGAAAACATG	16	66
AAAAURE	AAACAGAAAATATACAGAGATC CTCTGAACTTTTGTTTGTTAAT	20	59

template, a PCR product of approximately 1300 bp was obtained (Figure 14.3a, track 7). Sequencing of this product revealed the presence of an intron at position 520. These primers did not amplify a PCR product when the template genomic DNA was from mouse, hamster, or rat (Figure 14.3b, tracks 9, 10, 11). The human specificity of these primers made them suitable for screening the monochromosomal hybrid panel to assign p21.

The monochromosomal hybrid panel (Table 14.4) was screened with the p21-specific primer pair, A and B, by PCR. The 1300-bp PCR product was detected in the human controls (Figure 14.3b, tracks 12 and 13) and the hybrid MCP6BRA (Figure 14.3b, track 1). This hybrid contains a translocation of t(X,6) Xqter-q13:6p21-qter as its only human complement. As no PCR product was detected in the X-only hybrid, Hor19X (Figure 14.3b, track 8), p21 can be localized to 6p21-qter. A further translocation hybrid, 5647c122, was analyzed and produced a positive PCR product for p21 (Figure 14.3b, track 5). This hybrid contains the translocation of t (6:17) 6pter-6p21:17p13-qter. This localizes p21 to 6p21 (Figure 14.3c). Further localizations of p21 would be possible if the precise breakpoints in 6p21 in the hybrids MCP6BRA and 5647c122 were characterized.

3.   Radiation hybrids: telomeric assignment of four markers to chromosome 9p[29]

A more detailed description as to the methodologies involved in the construction and analysis of radiation hybrid (RH) panels can be found elsewhere.[12] A radiation hybrid panel was produced for chromosome 9 using the somatic cell hybrid GM10611 as the paternal cell line fused to the hamster cell A23. The hybrid GM10611 was previously shown by FISH and reverse painting onto spreads of normal human

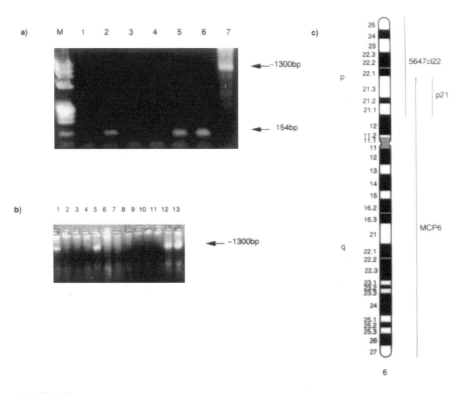

FIGURE 14.3

(a) Comparison of p21 cDNA and genomic PCR amplification products: The p21 154-bp cDNA PCR-amplified product was detected in HeLa (track 2), fetal retina (track 5), and normal breast (track 6) cDNA libraries. No product was PCR-amplified from the following cDNA libraries: mammary gland (track 1), lymphocyte (track 3), and breast adenocarcinoma (track 4). The p21 1300-bp genomic PCR-amplified product was detected in human genomic DNA (track 7). Phi-X 174 *Hae*III DNA fragments (track M) were used as molecular-weight-size markers. (b) PCR amplification of human p21: The p21 1300-bp PCR-amplified product was detected specifically in the somatic cell hybrids — MCP6BRA (track 1) and 5647c122 (track 5) — and the human control (tracks 12 and 13). No PCR product was detected in the somatic cell hybrids: GM07299 (track 2), GM10253 (track 3), GM10114 (track 4), Clone21E (track 6), DL18TS (track 7), and HORL9X (track 8), nor in mouse (track 9), hamster (track 10), and rat (track 11) genomic DNA. (c) Ideogram of human chromosome 6 showing the localization of p21/WAF1 to 6p21.

metaphase chromosomes to contain human chromosome 9 as the only cytogenetically detectable human material.[24]

A RH clone (H1) was selected from this panel and screened by PCR amplification using oligonucleotide primers specific for more than 40 single copy loci spanning chromosome 9. Two positive (chromosome 9-specific genomic DNA and total human genomic DNA) controls and one negative (hamster) control were used in all reactions. Only four loci (D9S54, D9S178, D9S288, and D9S132) amplified products on this clone. These all map in the interval 9pter–9p23. Primer sequences and general PCR reaction conditions for these four positive loci can be obtained from GDB via URL **http://gdbwww.gdb.org**. All markers tested on H1 distal to

(a)

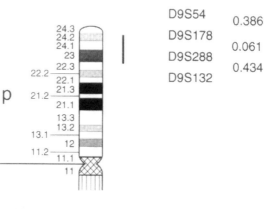

D9S54	0.386
D9S178	0.061
D9S288	0.434
D9S132	

(b)

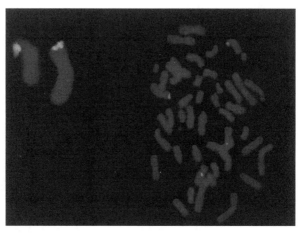

FIGURE 14.4

Correlation between cytogenetic and PCR data for RH clone H1: **(a)** Ideogram of human chromosome 9p showing the localization of the four positively screened PCR loci for RH clone H1 to 9pter–9p23. Distances between loci are expressed in cM. **(b)** DNA from RH clone H1 was used as a probe to paint normal human metaphase chromosomes. Only the telomere of 9p is fluorescently labeled (for clarification, the chromosome-9 pair has been enlarged).

D9S132 were negative (results not shown). Two-point and multipoint statistical analysis on a panel of 93 radiation hybrids, including H1, have ordered these four positive loci (Figure 14.4a) and confirmed their localization to 9pter–9p23 (results not shown). Additionally, further markers have been screened and integrated into this region and a radiation hybrid map constructed.[29] Confirmation of the chromosomal content of the RH clone H1 was determined by FISH and reverse painting (Figure 14.4b). This suggests that the RH clone H1 contains only a telomeric portion of 9p which can be used for assignment of any DNA fragments or genes to this chromosomal region.

IV. Summary

DNA-based gene mapping using somatic cell hybrids has risen exponentially over the last few years partly due to the development of large numbers of hybrids — more specifically, monochromosomal and radiation hybrid panels, but largely due to the discovery of the polymerase chain reaction (PCR). Southern blot methodology is still used for mapping DNA fragments, particularly when there is no DNA sequence available to design mapping primers. This method may also indicate whether a DNA fragment is part of a gene family or shows interspecies conservation. However, these latter two points often create problems in assignment — for example, in distinguishing human fragments from those of the rodent background.

The PCR technique has a number of advantages over Southern blotting for gene mapping:

1. Requires less template DNA
2. Much quicker with rapid "binning" of genes/random DNA fragments
3. Gene-specific (if follow primer design and PCR conditions detailed in methods), i.e., does not cross-hybridize with gene family members or across species

The majority of somatic cell hybrids (in particular, the monochromosomal hybrids) are well characterized in terms of chromosomal content, and most DNA sequences can be mapped with relative ease. However, problems in the assignment of certain genes do occur; either a gene is positive for two or more chromosomes, or, indeed, does not appear to map to any chromosome. This is probably due to deletions or translocations of small chromosomal regions in some hybrids. If this occurs, other hybrids should be tested or other gene-mapping techniques such as FISH should be considered for confirmation.

Radiation hybrids have a number of uses in gene mapping as well as regional assignment of DNA sequences. In particular, they provide a tool for the integration of genetic and physical maps by mapping nonpolymorphic DNA sequences, such as genes, relative to polymorphic DNA sequences. This is important in disease gene mapping studies.

In conclusion, DNA-based mapping using somatic cell hybrids has enabled the chromosomal assignment and in many cases the regional localization of thousands of genes and noncoding sequences. This resource is the starting point for many mapping studies.

Note: The monochromosomal hybrid DNA described in this chapter have been deposited at the U.K. HGMP Resource Center, Hinxton Hall, Hinxton, Cambridgeshire, CB10 1RQ, U.K., where 5- to 10-μg aliquots are available upon request to registered users of the center. Other sources of somatic cell hybrids are available from the literature.

References

1. Barski, G., Sorieul, S., and Cornefert, F., *C. R. Hebd. Seances Acad. Sci.,* 251, 1825, 1960.

2. Littlefield, J., *Science,* 145, 709, 1964.

3. Szybalski, E. H. and Szybalski, W., *Proc. Natl. Acad. Sci. USA,* 48, 2026, 1962.

4. Weiss, M. C. and Green, H., *Proc. Natl. Acad. Sci. USA,* 58, 1104, 1967.

5. Miller, D. J., Allderdice, P. W, Miller, D. W., Breg. W. R., and Migeon, B. R., *Science,* 173, 244, 1971.

6. Boone, C., Chen, T. R., and Ruddle, F. H., *Proc. Natl. Acad. Sci. USA,* 69, 510, 1972.

7. O'Brien, S. J., Simonson, J. M., and Eichelberger, M., in *Techniques in Somatic Cell Genetics,* Shaw, J.W., Ed., Plenum Press, New York, 1982.

8. Tunnacliffe, A. and Goodfellow, P., in *Genetic Analysis of the Cell Surface,* Goodfellow, P., Ed., Receptors and Recognition Series B16, Chapman and Hall, London, 1984, 57.

9. Deisseroth, A., Nienhuis, A., Turner, P., Velez, R., Anderson, F. W., Ruddle, F. H., Lawrence, J., Creagen, R., and Kucherlapati, R., *Cell,* 12, 205, 1977.

10. Deisseroth, A., Nienhuis, A., Lawrence, A., Giles, R., Turner, P., and Ruddle, F. H., *Proc. Natl. Acad. Sci. USA,* 75, 1456, 1978.

11. Cuticchia, A. J., in *Human Gene Mapping 1994, A Compendium,* Cuticchia, A. J., Ed., Johns Hopkins University Press, Baltimore, MD, 1995, 3.

12. Walter, P. and Goodfellow, P., *TIG,* 9, 73, 1991.

13. Abbott, C. and Povey, S., *Genomics,* 9, 73, 1991.

14. Theune, S., Fung, J., Todd, S., Sakaguchi, A. Y., and Naylor, S. L., *Genomics,* 9, 511, 1991.

15. Dubois, B. L. and Naylor, S. L., *Genomics,* 16, 315, 1993.

16. Drwinga, H. L., Toji, L. H., Kim, C. H., Greene, A. E., and Mulivor, R. A., *Genomics,* 16, 311, 1993.

17. Bobrow, M. and Cross, J., *Nature,* 251, 77, 1974.

18. Evans, H. J., Burkton, K. E., and Summer, A. T., *Chromosome,* 35, 320, 1971.

19. Gosden, J. R., in *Methods in Molecular Biology,* Pollard, J. W. and Walker, J. M., Eds., Humana Press, NJ, 1990.

20. Southern, E. M., *J. Mol. Biol.,* 98, 503, 1975.

21. Feinberg, A. P. and Vogelstein, B., *Anal. Biochem.,* 137, 266, 1984.

22. Saiki, R.K., Gelfand, D. H., Stoffel, S., Scharf, S. J., Higucchi, R., Horn, G. T., Mullis, K. B., and Erlich, H. A., *Science,* 239, 487, 1988.

23. Spurr, N. K., Kelsell, D., Rooke, L., Cavalli-Sforza, L. L., Bowcock, A., and Feder, F., *Ann. Hum. Genet.,* 55, 141, 1991.

24. Kelsell, D. P., Rooke, L., Warne, D., Bouzyk, M., Cullin, L., Cox, S., West, L., Povey, S., and Spurr, N. K., *Ann. Hum. Genet.,* 59, 233, 1995.

25. Polymeropoulos, M. H., Xiao, H., Glodek, A., Gorski, M., Adams, M. D., Moreno, R. F., Fitzgerald, M. G., Venter, J. C., and Merril, C. R., *Genomics,* 12, 492, 1992.

26. Polymeropoulos, M. H., Xiao, H., Sikela, J. M., Adams, M., Venter, J. C., and Merril, C. R., *Nature Genet.,* 4, 381, 1993.

27. Auffray, C., Béhar, G., Bois, F., Bouchier, C., DaSilva, C., Devignes, M. D. et al., *C. R. Acad. Sci. Paris (Life Sciences/Sciences de la Vie),* 318, 263, 1995.

28. Xiong, Y., Hannon, G. J., Zhang, H., Casso, D., Kobayashi, R., and Beach, D., *Nature,* 366, 701, 1993.

29. Bouzyk, M., Bryant, S. P., Cullin, L., Schmitt, K., Goodfellow, P. N., Ekong, R., Povey, S., and Spurr, N. K., *Genomics,* 34, 187, 1996.

30. Kelsall, A. and Meuth, M., *Somat. Cell Mol. Genet.,* 14, 149, 1988.

31. Dana, S. and Wasmuth, J. J., *Somatic Cell Genet.,* 8, 245, 1982.

32. Urlaub, G. and Chasin, L.A., *Proc. Natl. Acad. Sci. USA,* 77, 4216, 1980.

33. Creagan, R. P., Chen, S., and Ruddle, F. H., *Proc. Natl. Acad. Sci. USA,* 72, 2237, 1975.

34. Dana, S. and Wsmuth, J. J., *Mol.Cell Biol.,* 2, 1220, 1982.

35. Porterfield, B. W., Pomykala, H., Maltepe, E., Bohlander S. K., Rowley, J. D., and Diaz, M. O., *Somat. Cell Mol. Genet.,* 19, 460, 1993.

36. Chan, T.-S., Creagen, R. P., and Reardon, M. P., *Somat. Cell Genet.,* 4, 1, 1978.

37. Jones, G. E. and Sargent, P. A., *Cell,* 2, 43, 1974.

38. Littlefield, J. W., *Science,* 145, 709, 1964.

39. Ayusama, D., Koyama, H., Iwata, K., and Seno, T., *Somatic Cell Genet.,* 7, 523, 1981.

40. Cirullo, L. R., Arrendondo-Vega, F. X., Smith, M., and Wasmuth, J. J., *Somat. Cell Genet.,* 9, 215, 1983.

41. Sawami, H., Ito, K., Norioka, M., Monden, S., Fujita, M., and Uchino, H., *Nippon Ketsueki Gakkai Zasshi,* 52, 1033, 1989.

42. Littlefield, J. W., *Exp. Cell Res.,* 41, 190, 1966.

43. Colbere-Garapin, F., Horodniceaun, F., Kourilsky, P., and Garapin, A., *J. Mol. Biol.,* 150, 1, 1981.

44. Mansour, S. L., Thomas, K. R., and Capecchi, M.R., *Nature,* 336, 348, 1988.

45. Mulligan, R. C. and Berg, P., *Proc. Natl. Acad. Sci. USA,* 78, 2072, 1981.

46. Gritz, L. and Davies, J., *Gene,* 25, 179, 1983.

47. Vara, J. A., Portela, A., Ortin, J., and Jimenez, A., *Nucl. Acids Res.,* 14, 4617, 1986.

48. Hartman, S. C. and Mulligan, R. C., *Proc. Natl. Acad. Sci. USA,* 85, 8047, 1988.

49. Morgenstern, J. P. and Land, H., *Nucl. Acids Res.,* 18, 3587, 1990.

50. Eglitis, M. A., *Human Gene Ther.,* 2, 195, 1991.

51. Cartier, M., Chang, M. W. M., and Stanners, C.P., *Mol. Cell. Biol.,* 7, 1623, 1987.

52. O'Hare, K., Benoist, C., and Breathnach, R., *Proc. Natl. Acad. Sci. USA,* 78, 1527, 1981.

Index

Index